Earthquake-Compendium

Shailendra Tripathi

Banshidhar Pandey

Published by

BONFRING®
Intellectual Integrity

Earthquake-Compendium

ISBN 978-93-85477-66-9

Author

Shailendra Tripathi

Banshidhar Pandey

Bonfring
309, 2nd Floor, 5th Street Extension, Gandhipuram,
Coimbatore-641 012.
Tamilnadu, India.
E-mail: info@bonfring.org
Website: www.bonfring.org
Phone: 0422-3928700

Disclaimer

Subject matters and technical contents of the book **"Earthquake-Compendium"** are principally based on the published literatures-References have been cited and duly acknowledged without any claim of originality. Authors have sincerely edited/restructured the text, integrated together the subject matters, and made them coherent purely with the objectives of **"Understanding the Science behind Earthquakes"** as well as **"Knowledge and Information Dissemination"** to suit the wide spectrum of readers right from the common ones up to the level of specialists.

Dedication

The Book "Earthquake-Compendium" is dedicated to Mr. R P Tewary beloved father of Mr. Shailendra Pati Tripathi –a respected Senior Citizen and a great devotee of Nature. He believes in Supreme Power of Nature. Only because of his blessings, best wishes and continued encouragements, it has been possible to write this book for the benefits of wide spectrum of readers. I and Shailendra dedicate this Book to him with love and honor.

Preface

A **natural disaster** is a major adverse event resulting from natural processes of the Earth; examples include floods, volcanic eruptions, earthquakes, tsunamis, and other geologic processes. A natural disaster can cause loss of life or property damage, and typically leaves some economic damage in its wake, the severity of which depends on the affected population's resilience, or ability to recover.

Frequent outbreaks of natural disasters are a serious problem on a global scale. An adverse event will not rise to the level of a disaster if it occurs in an area without vulnerable population. In a vulnerable area, however, such as San Francisco and Nepal, an earthquake can have disastrous consequences and leave lasting damage, requiring years to repair.

An earthquake is ground shaking phenomena caused by a sudden movement of rock in the earth's crust. Such movements occur along faults, which are thin zones of crushed rock separating blocks of crust. When one block suddenly slips and moves relative to the other along a fault, the energy released creates vibrations called seismic waves that radiate up through the crust to the earth's surface, causing the ground to shake.

As far as the cause of earthquake is concerned, it is the result of a sudden release of energy in the Earth's crust that creates seismic waves. At the Earth's surface, earthquakes manifest themselves by vibration, shaking and sometimes displacement of the ground. Earthquakes are caused mostly by slippage within geological faults, but also by other events such as volcanic activity, landslides, mine blasts, and nuclear tests. The underground point of origin of the earthquake is called the *focus*. The point directly above the focus on the surface is called the ***epicenter***. Earthquakes by themselves rarely kill people or wildlife. It is usually the secondary events that they trigger, such as building collapse, fires, tsunamis (seismic sea waves) and volcanoes. These secondary events are actually the cause of human disaster. Many of these could possibly be avoided by better construction, safety systems, early warning and planning.

Some 80 percent of all the planet's earthquakes occur along the rim of the Pacific Ocean, called the "Ring of Fire" because of the preponderance of volcanic activity there as well. Most earthquakes occur at fault zones, where tectonic plates—giant rock slabs that make up the Earth's upper layer—collide or slide against each other. These impacts are usually gradual and unnoticeable on the surface; however, immense stress can build up between plates. When this stress is released quickly, it sends massive vibrations, called seismic waves, often hundreds of miles through the rock and up to the surface. Other quakes can occur far from faults zones when plates are stretched or squeezed.

Scientists assign a magnitude rating to earthquakes based on the strength and duration of their seismic waves. A quake measuring 3 to 5 is considered minor or light; 5 to 7 is moderate to strong; 7 to 8 is major; and 8 or more is the largest.

Intense ground shaking can generate many sources of potential harm or loss. In the natural environment, such hazards include the following:

- **Landslides** or avalanches.
- **Surface faulting**, in which the surface of the ground along one side of a fault is displaced horizontally or vertically in relation to the ground on the other side.
- **Tsunamis**, which can be triggered by earthquake-induced underwater landslides or by surface faulting that occurs on the floor of the ocean.
- **Liquefaction**, in which loosely packed, water-logged soils temporarily lose strength and stiffness and behave like liquids, causing the ground to sink or slide.
- **Flash floods**, which can be caused by liquefaction near rivers or lakes.

These hazards, as well as the ground shaking that may produce them, can also create a variety of hazards in the built environment. Buildings—or their components or contents—can be collapsed, toppled, broken apart, tossed around or rendered inoperable or unusable. The same can happen to lifeline infrastructure systems and their components, including those related to transportation, such as roads, bridges, railways, ports and airports and those related to utilities, such as distribution lines for water, wastewater, electric power, telecommunications, natural gas and liquid fuels.

Damage incurred among these hazards, such as broken gas or water pipes, can itself be hazardous, generating further damage by igniting fires or flooding buildings.

It is therefore, important and essential to remain prepared in advance for such disasters in order to minimize the number of casualties. All possible measures must be applied to raise the survival rate, and the contribution of advanced technologies such as robotics is expected in the future.

Scientific aspects of the earthquakes contained in this particular compendium would go a long way in understanding the science behind the earthquakes and mitigating the resultant sufferings. **Further**, this compendium could be a very good reference book for the scientific and technical facts related to earthquakes.

Acknowledgement

Subject matters and technical contents of the Book—Earthquake Compendium are principally based on the published literatures available on the Internet Websites. Authors acknowledge with regards all those whose publications have been referred and used in preparation of the book

Author's Profile

Shailendra Tripathi,
National WHS Manager and Head,
QBE,
Australia.

Shailendra Tripathi is an experienced Work Health and Safety professional currently heading the WHS portfolio at QBE Insurance group, Australia. Shailendra has a degree in Materials Engineering and has worked with organization like Tata Steel, Lloyds in India, Smorgon, Wesfarmers , Alstom and Roads and Traffic Authority in Australia. Shailendra has expertise in Quality, Safety and Environmental systems and exposure to one of the best technologies around the world. Shailendra has worked in senior management roles within Rail, Construction, High volume manufacturing and Iron and Steel production environment. Shailendra is an active conference speaker on safety, human factor, resilience, quality and environment. He has more than 20 years of practical experience in the field of safety, quality and environment.

Competencies

- Strategic Planning, developing, documenting, implementing and review of Quality, Environment, Health and Safety and Risk Management solutions;
- Certified Quality (ISO 9001), Environment (ISO 14001) and Occupational Health and Safety (AS/NZS 4801) system auditor;
- Professional member of the *TMS* international.
- Board member of BSI certification advisory committee
- Member of Leadership Consortium

Shailendra hobbies include growing vegetables, coaching cricket, reading professional journals and addressing conferences.

Banshidhar Pandey,

Retired Researcher,

R & D, Tata Steel,

Jamashedur,

India

B D Pandey, Retired Researcher, R&D Tata Steel, Jamashedur, India - Mr. B D Pandey Graduated in Metallurgical Engineering in1967, obtained degree of M. Tech (Ferrous Metallurgy) in 1981 from NIT (then RIT) Jamshedpur. He joined Tata Steel (R&D) in June 1970 and retired as Researcher in February 2004 after serving Tata Steel for 34 years. As a Researcher he headed the Iron-making process Research in R & D. His areas of specialization are Iron-Making, Direct Reduction, Value added Ferroalloys. To his credit, Mr. Pandey has about 100 publications in the National and International Journals of repute. He is the Recipient of S. Vishwanathan Medal (1994) for the best paper in Tata Search, Visvesvaraya Gold Medal (1996-1997), SAIL Gold Medal (1997-1998). He is the Co-author of an Engineering Book "Metallic for Steel making- production and use". Mr. Pandey enjoys publishing his research findings. In BYBF (Beyond the Blast Furnace) another engineering book published by the CRC Press America, Mr., Pandey contributed substantially through his research findings. Mr. Pandey loves Nature. Natural phenomena / events are very close to his heart. Besides applied research, Mr. Pandey worked in the areas of Business Process Re-engineering (Total Quality Management, Knowledge Management – lead expert, Total Operational Performance, Environment and Safety).

Chapter	**Contents**	**Page No**

ABSTRACT

Earthquake Facts

An earthquake is the result of a sudden release of energy in the Earth's crust that creates seismic waves. At the earth's surface, earthquakes manifest themselves by vibration, shaking and sometimes displacement of the ground. Earthquakes are mostly caused by slippage within geological faults. It can also be caused by other events such as volcanic activity, landslides, mine blasts, and nuclear tests

Earthquakes by themselves rarely kill people or wildlife. It is usually the secondary events that they trigger, such as building collapse, fires, tsunamis (seismic sea waves) and volcanoes, which actually cause the human disaster.

It is to be mentioned that about 80 percent of all the planet's earthquakes occur along the rim of the Pacific Ocean, called the "Ring of Fire" because of the preponderance of volcanic activity there as well. Most earthquakes occur at fault zones, where tectonic plates-giant rock slabs that make up the Earth's upper layer-collide or slide against each other. These impacts are usually gradual and unnoticeable on the surface; however, immense stress can build up between plates. When this stress is released quickly, it sends massive vibrations, called seismic waves, often hundreds of miles through the rock and up to the surface. Other quakes can occur far from faults zones when plates are stretched or squeezed.

On average, a magnitude 8 quake strikes somewhere every year and some 10,000 people die in earthquakes annually. Collapsing buildings claim by far the majority of lives, but the destruction is often compounded by mud slides, fires, floods, or tsunamis. Smaller temblors that usually occur in the days following a large earthquake can complicate rescue efforts and cause further death and destruction.

Loss of life and property can be avoided through emergency planning, education, early warning, better safety systems and improved construction design of buildings that sway rather than break under the stress of an earthquake.

Definition

An **earthquake** is the perceptible shaking of the surface of the earth, which can be violent enough to destroy major buildings and kill thousands of people. When a rock below earth's surface breaks the earthquake takes place due to the sudden release of energy in the Earth's crust that creates seismic waves.

Energy builds up underground and once it builds up to the maximum limit, it just can't hold back any longer, and then there is an explosive release of energy. When the energy is released it radiates outward in all directions. As the energy travels toward earth's surface from underground, it shakes the ground, sometimes so much that it can cause damage and destructions.

Stress concentration along the plate boundaries on the earth surface is the main cause of the earthquake. The plates are dynamic, so they are always moving. Sometimes they move to such an extent that they push into each other or pull apart. **Compression stress** occurs when rocks are pushed together - they're pressed into one another. **Tensional stress** occurs when rocks are pulled apart - they're being stretched farther than they would be otherwise. **Shear stress** is when rocks slide past each other in opposite directions.

Components

When the stress becomes high enough, the rock breaks and the ground begins to shake. If the rock splits into separate pieces, a *fault* is created which is the line of fracture along the split rock. Depending on the relative layout of the rock blocks that move relative to each other in an earthquake and the direction of their relative movements, there are several types of faults. Some of the most common types of faults are: normal fault, reverse fault and strike-slip fault.

According to elastic rebound theory, stresses build on both sides of a fault, causing the rocks to deform plastically. When the stresses become high enough, the rocks return to their original shape but they move. This motion releases the energy that causes an earthquake.

It is to be understood that not all of the breaking of rock and energy release takes place at once during an earthquake. There may be *foreshocks* and *aftershocks,* which are the energy released before and after the main quake.

The point underground where the actual breaking of the rock occurs is called the *focus*. It is actually the focal point of the earthquake. This is where the main event occurs underground. The point directly above the focus on the surface of Earth is called the *epicenter.* This is where the ground shaking is usually the strongest. From this point on the surface, the waves of energy from below ground begin to travel outward, so this is considered as the central point of shaking above ground. Because the shaking is strongest here, this is also where the most damage usually occurs.

Earthquakes can cause immense loss to the lives and property .Buildings crumble and fall, landslides and avalanches may be triggered, and roads and bridges can collapse. This is considered as one of the worst Natural calamities

The Structure of the Earth Tectonic Plates

The Earth consists of four concentric layers: inner core, outer core, *mantle* and *crust*. The crust is made up of tectonic plates, which are in constant motion. Earthquakes and volcanoes are most likely to occur at plate boundaries.

The four distinct concentric layers are:

1. **The inner core** is in the centre and is the hottest part of the Earth. It is solid and made up of iron and nickel with temperatures of up to 5,500°C. With its immense heat energy, the inner core is like the engine room of the Earth.
2. **The outer core** is the layer surrounding the inner core. It is a liquid layer, also made up of iron and nickel. It is still extremely hot, with temperatures similar to the inner core.
3. **The mantle** is the widest section of the Earth. It has a thickness of approximately 2,900 km. The mantle is made up of semi-molten rock called magma. In the upper parts of the mantle the rock is hard, but lower down the rock is soft and beginning to melt.
4. **The crust** is the outer layer of the earth. It is a thin layer between 0-60 km thick. The crust is the solid rock layer upon which we all live. There are two different types of crust: **continental crust**, which carries land, and **oceanic crust**, which carries water.

Seismic discontinuities aid in distinguishing divisions of the Earth into inner core, outer core, D" layer, lower mantle, transition region, upper mantle, and crust (oceanic and continental).

- **Inner core: 1.7% of the Earth's mass; depth of 5,150-6,370 kilometers.** The inner core is solid and unattached to the mantle, suspended in the molten outer core. It is believed to have solidified as a result of pressure-freezing which occurs to most liquids when temperature decreases or pressure increases.
- **Outer core: 30.8% of Earth's mass; depth of 2,890-5,150 kilometers.** The outer core is a hot, electrically conducting liquid within which convective motion occurs. This conductive layer combines with Earth's rotation to create a dynamo effect that maintains a system of electrical currents known as the Earth's magnetic field. It is also

responsible for the subtle jerking of Earth's rotation. This layer is not as dense as pure molten iron, which indicates the presence of lighter elements. Scientists suspect that about 10% of the layer is composed of sulfur and/or oxygen because these elements are abundant in the cosmos and dissolve readily in molten iron.

- **D" layer: 3% of Earth's mass; depth of 2,700-2,890 kilometers.** This layer is 200 to 300 kilometers thick and represents about 4% of the mantle-crust mass. Although it is often identified as part of the lower mantle, seismic discontinuities suggest the D" layer might differ chemically from the lower mantle lying above it. Scientists theorize that the material either dissolved in the core, or was able to sink through the mantle but not into the core because of its density.

- **Lower mantle: 49.2% of Earth's mass; depth of 650-2,890 kilometers.** The lower mantle contains 72.9% of the mantle-crust mass and is probably composed mainly of silicon, magnesium, and oxygen. It probably also contains some iron, calcium, and aluminum. Scientists make these deductions by assuming the Earth has a similar abundance and proportion of cosmic elements as found in the Sun and primitive meteorites.

- **Transition region: 7.5% of Earth's mass; depth of 400-650 kilometers.** The transition region or mesosphere (for middle mantle), sometimes called the fertile layer, contains 11.1% of the mantle-crust mass and is the source of basaltic magmas. It also contains calcium, aluminum, and garnet, which is a complex aluminum-bearing silicate mineral. This layer is dense when cold because of the garnet. It is buoyant when hot because these minerals melt easily to form basalt which can then rise through the upper layers as magma.

- **Upper mantle: 10.3% of Earth's mass; depth of 10-400 kilometers.** The upper mantle contains 15.3% of the mantle-crust mass. Fragments have been excavated for our observation by eroded mountain belts and volcanic eruptions. Olivine $(Mg,Fe)_2SiO_4$ and pyroxene $(Mg,Fe)SiO_3$ have been the primary minerals found in this way. These and other minerals are refractory and crystalline at high temperatures; therefore, most settle out of rising magma, either forming new crustal material or never leaving the mantle. Part of the upper mantle called the asthenosphere might be partially molten.

- **Oceanic crust: 0.099% of Earth's mass; depth of 0-10 kilometers.** The oceanic crust contains 0.147% of the mantle-crust mass. The majority of the Earth's crust was made through volcanic activity. The oceanic ridge system, a 40,000-kilometer network of volcanoes, generates new oceanic crust at the rate of 17 km^3 per year, covering the ocean floor with basalt. Hawaii and Iceland are two examples of the accumulation of basalt piles.

- **Continental crust: 0.374% of Earth's mass; depth of 0-50 kilometers.** The continental crust contains 0.554% of the mantle-crust mass. This is the outer part of the Earth composed essentially of crystalline rocks. These are low-density buoyant minerals dominated mostly by quartz (SiO2) and feldspars (metal-poor silicates). The crust (both oceanic and continental) is the surface of the Earth; as such, it is the coldest part of our planet. Because cold rocks deform slowly, we refer to this rigid outer shell as the lithosphere (the rocky or strong layer).

Salient Features

- Earthquakes are the phenomena experienced during sudden movements of the Earth's crust. Under the Earth's crust lies the asthenosphere, the upper part of the mantle composed of liquid rock. The plates of the Earth's crust essentially "float" on top of this layer, and can be forced to shift as the upwelling molten material below moves. As the plates shift (and thus interact with each other), an enormous amount of energy is released in the form of waves. Although earthquakes can occur anywhere on the planet with little or no warning, the most extreme earthquakes occur near plate boundaries, as the plates converge (collide), diverge (move away from another), or shear (grind past one another).

- Moving rock and magma within volcanoes can also trigger earthquakes. In all of these cases, large sections of the crust can fracture and move to-and-fro to dissipate the released energy. This "shaking" is the sensation felt during an earthquake. The energy released is often described in terms of "magnitude," a logarithmic scale used to describe how energetic an earthquake was; a quake of magnitude 2 is hardly noticeable without special monitoring equipment, while quakes over magnitude 8 may actually cause the ground to visibly heave and roll. Since the scale is logarithmic, a magnitude 8 quake is not four times more energetic than a magnitude 2 quake, but one billion times more energetic!

Relevant Characteristics

Earthquakes occur when vast plates, or rocks, within the earth suddenly break or shift under stress, sending shock waves rippling.

- Sudden movement along the fault causes the ground to shake.
- Most earthquakes occur along fractures in the Earth's crust called faults.
- Intra-plate quakes occur far from plate edges and happen when stress builds up and the Earth's crust is stretched or squeezed together until it rips.
- A major *earthquake* is usually rather *short in duration*, often *lasting only a few seconds* and seldom more than a minute or so.
- There are several different types of faults. Each can be a few inches or many hundreds of miles long. They can be horizontal, vertical, or at an angle.
- Earthquake waves are measured on sensitive instruments called seismographs.
- The Richter scale assigns quakes a number based on the power of its seismic waves.
- Thousands of quakes occur every day around the globe, most of them too weak to be felt.
- Every year about 10,000 people, on average, die as a result of earthquakes.

Seismic Waves

The energy from earthquakes travels in waves called **seismic waves**. These seismic waves are used to find out the details about earthquakes and also about the Earth's interior.

Seismic waves move outward in all directions away from their source. There are two major types of seismic waves viz., **body waves** and **surface waves** .Body waves travel through the solid body of the Earth from the earthquake's focus throughout the Earth's interior and to the surface. Surface waves just travel along the ground surface. The different types of seismic waves travel at different speeds in different materials. All seismic waves travel through rock, but not all travel through liquid or gas. In an earthquake, body waves are responsible for sharp jolts. Surface waves are responsible for rolling motions. Surface waves do most of the damage in an earthquake.

Tsunami

Earthquakes can cause deadly ocean waves called **tsunami**. When ocean water is displaced by the sharp jolt of an undersea earthquake, the seismic energy forms a set of waves. The waves travel through the sea entirely unnoticed since they have low amplitudes and long wavelengths. When these waves come onto shore, they can grow to enormous heights and cause tremendous destruction and loss of life.

Volcanic Eruption

The process that leads to volcanic eruptions consists of the following steps:

- Swarms of small earthquakes precede eruptions
- The frequency of the earthquakes increases
- They reach a point of such rapid succession that they create a harmonic tremor
- The frequency of the harmonic tremor steadily rises
- The volcano 'screams' due to incredible pressure building up
- The tremor and the screaming stops
- The volcano erupts

General Effects of Earthquakes

Earthquakes are natural disasters of a generally unpredictable nature. In spite of considerable efforts made towards improving the understanding of these natural disasters and protecting built environment from their effects, earthquakes still cause huge human and economic losses; this is true both for highly industrialized and lesser developed countries.

A general study of earthquakes includes: consideration of the nature of ground faults, the propagation of shock waves through the earth mass, the specific nature of recorded major quakes, etc.

The ground movements caused by earthquakes can have several types of damaging effects. Some of the major effects are:

1. Ground shaking, i.e. back-and-forth motion of the ground, caused by the passing waves of vibration through the ground;
2. Soil failures, such as liquefaction and landslides, caused by shaking;
3. Surface fault ruptures, such as cracks, vertical shifts, general settlement of an area, landslides, etc.
4. Tidal waves (tsunamis), i.e. large waves on the surface of bodies of water that can cause major damage to shoreline areas.

Shaking and ground rupture are the main effects created by earthquakes, principally resulting in more or less severe damage to buildings and other rigid structures.

April 2015 Nepal Earthquake

In order to illustrate devastations caused due to an earthquake, a typical earthquake that occurred in Nepal on 25 April 2015 has been described in detail.

Nepal Earthquake is one of the deadliest earthquakes which took place in Nepal due to continental collision of Indian and Eurasion plates. As the Indian subcontinent pushes against Eurasia, pressure is released in the form of earthquakes. The constant crashing of the two plates forms the Himalayan mountain range. During this particular earthquake a chunk of rock about 9 miles (15 Kilometers) below the earth's surface shifted, unleashing a shock wave (described as being as powerful as the explosion of more than 20 thermonuclear weapons) that ripped through the Katmandu Valley. The reason is the regular movement of the fault line that runs along Nepal's southern border, where the Indian subcontinent collided with the Eurasia plate 40 million to 50 million years ago.

The intensity of the earthquake reported was 9.0 on the Richter magnitude scale. In this earthquake, 9000 persons were killed, 22,000 got injured and the loss of properties amounted to $ 5 billion (equivalent to 25% of Nepal's GDP). It was the worst natural disaster to strike Nepal since the 1934 Nepal–Bihar earthquake.

The compendium consists of seven sections covering important aspects of Earthquakes as listed below:

- **Section 1**-Introduction
- **Section2**-Earthquake characterization
- **Section3**-Inner geological structure of the earth
- **Section4**-Science of earthquakes (Seismology and Seismic waves)
- **Section5**-Earthquake effects
- **Section6**- Earthquakes and Tsunamis in India- 2015 Nepal earthquakes
- **Section7**-Images Illustrating Characteristic Features of the Earthquakes

Besides knowledge and information dissemination, this particular compendium is expected to enhance the knowledge and understanding of the science behind earthquakes. The contribution of advanced technologies such as robotics is expected in the future for better understandings of the earthquakes, predicting the earthquakes and assessing the damages and destructions caused.

Aim/ Objective

- Understanding the Science Behind Earthquakes
- Knowledge and Information Dissemination

Synopsis

An earthquake is the result of a sudden release of energy in the Earth's crust that creates seismic waves. At the earth's surface, earthquakes manifest themselves by vibration, shaking and sometimes displacement of the ground. Earthquakes are mostly caused by slippage within geological faults. It can also be caused by other events such as volcanic activity, landslides, mine blasts, and nuclear tests.

In an earthquake, the initial movement that causes seismic vibrations occurs when two sides of a fault suddenly slide past each other. The size of an earthquake depends on:

- The area of the fault that ruptured.
- The distance that the rocks on the two sides of the fault slide past one another.

Small earthquakes rupture small faults or small sections of large faults. Fault movement during such events is quick- small quakes last only a fraction of a second and the rocks on either side of the fault don't move far away.

Large earthquakes rupture faults that are tens to thousands of kilometers long. Such ruptures can take minutes to complete, so strong shaking near the earthquakes can last several minutes and rocks across the fault can be offset tens of meters during very large earthquakes.

The most commonly used quantification of earthquake size is the magnitude-the amplitude of ground shaking that is measured by an instrument called seismograph. Large earthquakes are less frequent than smaller ones. The temporal distribution of earthquakes by size follows a logarithmic rule.

Earthquakes are not isolated events, they occur in sequences. Most often, each sequence is dominated by an event with a larger magnitude than all others in the sequence. The largest event is called the main shock, and the events that follow are called aftershocks. Occasionally, the main shock is preceded by an event or events that are called the foreshock(s). Sometimes, earthquakes occur in interesting sequences which are called doublets, triplets, multiplets or swarms depending on how many similar-size events are in the sequence.

S P Tripathi: Head, WHS, QBE Australia

B D Pandey: Retired Researcher, Tata Steel, Jamshedpur, India

Subject matters and technical contents of the book **"Earthquakes-Compendium"** are principally based on the published literatures-References have been cited and duly acknowledged

Most earthquakes occur at fault zones, where tectonic plates—giant rock slabs that make up the Earth's upper layer—collide or slide against each other. These impacts are usually gradual and unnoticeable on the surface; however, immense stress can build up between plates. When this stress is released quickly, it sends massive vibrations, called seismic waves, often hundreds of miles through the rock and up to the surface. Other quakes can occur far from faults zones when plates are stretched or squeezed

The compendium consists of seven sections covering important aspects of Earthquakes. Besides knowledge and information dissemination, this particular compendium is expected to enhance the knowledge and understanding of the science behind earthquakes

1. Introduction

Earthquake (also known as a quake, *tremor* or temblor) is the perceptible shaking of the surface of the Earth, which can be violent enough to destroy major buildings and kill thousands of people. The severity of the shaking can range from barely felt to violent enough to toss people around. Earthquakes result from the sudden release of energy in the Earth's crust that creates seismic waves. The seismicity, seismism or seismic activity of an area refers to the frequency, type and size of earthquakes experienced over a period of time.

Earthquakes are measured using observations from seismometers. The Moment Magnitude is the most common scale on which earthquakes larger than approximately 5 are reported for the entire globe. The more numerous earthquakes smaller than magnitude 5 reported by national seismological observatories are measured mostly on the local magnitude scale, also referred to as the Richter magnitude scale. These two scales are numerically similar over their range of validity. Magnitude 3 or lower earthquakes are mostly almost imperceptible or weak and magnitude7 and over potentially cause serious damage over larger areas, depending on their depth. The largest earthquakes in historic times have been of magnitude slightly over 9, although there is no limit to the possible magnitude. As of March 2014, the most recent large earthquake of magnitude 9.0 or larger was a 9.0 magnitude earthquake in Japan in 2011. It was also the largest Japanese earthquake since records began. Intensity of shaking is measured on the modified Mercalli intensity scale. All else being equal, [1] the shallower an earthquake, the more damage to structures and lives it causes.

At the Earth's surface, earthquakes manifest themselves by shaking and sometimes displacement of the ground. When the epicenter of a large earthquake is located offshore, the sea bed may be displaced sufficiently to cause a tsunami. Earthquakes can also trigger landslides, and occasionally volcanic activity.

In its most general sense, the word ***earthquake*** is used to describe any seismic event-whether natural or caused by humans-that generates seismic waves. Earthquakes are caused mostly by rupture of geological faults, but also by other events such as volcanic activity, landslides, mine blasts, and nuclear tests. An earthquake's point of initial rupture is called its focus or hypocenter. The epicenter is the point at ground level directly above the hypocenter.

Earthquakes, also called temblors, can be so tremendously destructive it's hard to imagine. The Great Sumatra Earthquakes and Indian Ocean Tsunamis of 26 December 2004 and 28 March 2005 together with Nepal Earthquake that occurred on 25 April 2015 amply demonstrate the devastations caused due to these two natural calamities:

The Great Sumatra Earthquake: At 7:58 A.M. local time on Sunday, 26 December 2004, a great earthquake occurred in the Indian Ocean approximately 250 km west of Sumatra, Indonesia. With a moment magnitude of 9.1–9.3, this was the second largest instrumentally recorded earthquake in history. The earthquake had average source duration of about 500 seconds, the longest ever recorded, and a rupture length of about 1,300 km, the largest ever determined instrumentally. This earthquake generated one of the most devastating tsunamis in recorded history. The tsunami of December 26 affected at least 16 nations directly, and indirectly affected the entire globe. The initial impact of the tsunami was the inundation of the cities of Banda Aceh and Meulaboh along the northwestern coast of the island of Sumatra. There were 131,000 confirmed deaths in this region and 37,000 people were listed as missing. Over 80,000 houses sustained damage or collapsed, and it is estimated that 518,000 people were displaced from their homes. Within hours the tsunami brought death and destruction to the shores of Thailand to the east and to Sri Lanka, India, and the Maldives to the west. The tsunami also caused death and destruction in Somalia and other nations along the east coast of Africa. The death toll from the tsunami has been estimated to be about 300,000 - this includes local residents of the affected regions as well as a sizeable number of tourists from Europe, Asia, and the Americas. In addition to the direct loss of life was the extensive damage to homes, hospitals, schools, centre of worship, government buildings, infrastructure, and the social fabric of the affected communities. It will be years before the affected regions will return to anything like a "normal" existence. Three months after the December earthquake, on 28 March 2005, another large earthquake with a moment magnitude of about 8.7 struck off the coast of Sumatra near the island of Nias. This earthquake generated only a rather small tsunami. However, approximately 900 people died and 22,000 people were displaced by the event.

Earthquakes and Tsunamis in India: The Indian subcontinent has a history of earthquakes. The reason for the high frequency and intensity of earthquakes is the Indian plate driving into Asia at a rate of approximately 49 mm/year. As a result one should constantly remain on alert. If the main centre exists in Himalaya range then it will damage maximum part of northern India. It is to be noted that the Himalayas being a fairly young mountain range is undergoing constant geological changes.

India, due to its, physiographic and climatic conditions is one of the most disaster prone areas of the world. It is vulnerable to windstorms from both the Arabian Sea and Bay of Bengal. There are active crustal movements in the Himalaya leading to earthquakes. About 58.7 % of the total land mass is prone to earthquake of moderate to very high intensity. Further, India has increasingly become vulnerable to Tsunamis since the 2004 Indian Ocean Tsunami. It is important to mention that India has a coastline running 7600 km long; as a result it is repeatedly threatened by cyclones. India is also very prone to earthquakes as well. The major reason for the high frequency and intensity of the earthquakes is that the Indian plate is driving into Asia at a rate of approximately 47 mm/year. As per the Geographical statistics, almost 54% of the land in India is vulnerable to earthquakes. According to the estimates shown by a World Bank and United Nations report; around 200 million city dwellers in India will be exposed to storms and earthquakes by 2050.

2015 Nepal Earthquake: Nepal earthquake that occurred on 25 April 2015 is one of the deadliest earthquakes which took place in Nepal due to continental collision of Indian and Eurasion plates. As the Indian subcontinent pushes against Eurasia, pressure is released in the form of earthquakes. The constant crashing of the two plates forms the Himalayan mountain range. During this particular earthquake a chunk of rock about 9 miles (15 Kilometers) below the earth's surface shifted, unleashing a shock wave (described as being as powerful as the explosion of more than 20 thermonuclear weapons) that ripped through the Katmandu Valley. The reason is the regular movement of the fault line that runs along Nepal's southern border, where the Indian subcontinent collided with the Eurasia plate 40 million to 50 million years ago. The intensity of the earthquake reported was 9.0 on the Richter magnitude scale. In this earthquake, 9000 persons were killed, 22,000 got injured and the loss of properties amounted to $ 5 billion (equivalent to 25% of Nepal's GDP). It was the worst natural disaster to strike Nepal since the 1934 Nepal–Bihar earthquake.

Owing to the massive loss of lives and properties caused due to major earthquakes, it is important and essential to be prepared for such disasters in order to minimize the number of casualties and huge economic losses.

Reference

1. "Earthquake FAQ". Crustal.ucsb.edu. Retrieved2011-07-24.

2. Earthquake Characterization

Abstract

Important parameters for characterizing earthquakes are described here. These include:

- Events during the earthquake
- Earthquake zones
- Earthquake theory
- Ground shaking- acceleration of the ground in one horizontal direction plotted as a function of elapsed time (Acceleration –Time records)
- Magnitude, intensity and classes.

The most widely accepted indicators of the size of an earthquake are its magnitude and intensity. The magnitude is a measure of an earthquake in terms of the released energy. At the present time, the most popular scale is the Richter scale. *Modified Mercalli Intensity (MMI) Scale is currently used in United States. Earthquakes with an intensity of 8 or more on Richter scale and (VIII) or more on MMI scale can totally destroy communities and structures near the* epicenter.

2.1. Earthquake Zones [1]

Some locations are prone to earthquakes and some are not. Nearly 95% of all earthquakes take place along one of the three types of plate boundaries-divergent, convergent and transform.

The region of the planet with the most earthquakes is the area around the Pacific Ocean. About 80% of all earthquakes strike this area. This region is called the Pacific Ring of Fire because most volcanic eruptions occur there as well. The Pacific Ocean is surrounded by convergent and transform plate boundaries.

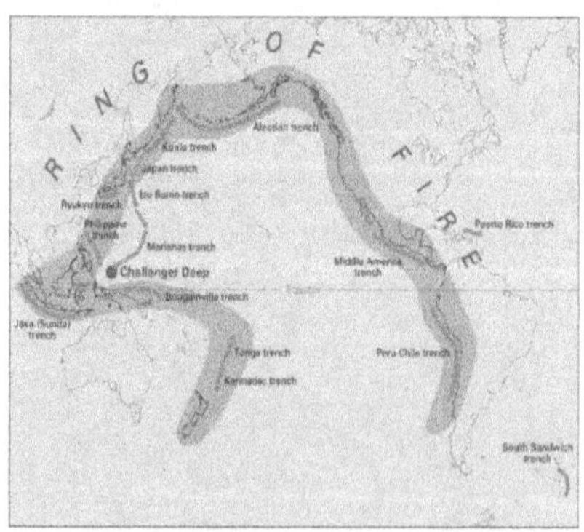

Figure: The Pacific Ring of Fire is the most geologically active region of the world. Convergent plate boundaries cause earthquakes and volcanic eruptions all around the Pacific Ocean basin. There are also transform plate boundaries along the San Andreas Fault in California and the Alpine Fault in New Zealand.

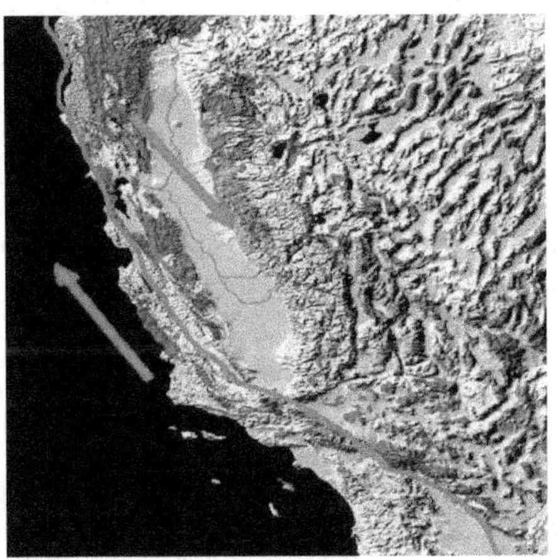

Figure: The San Andreas Fault runs up western California. It is the transform plate boundary between the northwestward moving Pacific Plate and the southeastward moving North American Plate.

About 15% of all earthquakes take place in the Mediterranean-Asiatic belt. This is where convergent plate boundaries are shrinking the Mediterranean Sea and causing the Himalayas to grow. The remaining 5% of earthquakes are scattered around the other plate boundaries with a few occurring in the middle of a plate, away from plate boundaries.

All three types of plate boundaries have earthquakes. Enormous and deadly earthquakes occur at transform plate boundaries. Because the slabs of lithosphere slide past each other without moving up or down, transform faults have shallow focus earthquakes. The most notorious earthquake fault in North America is the San Andreas Fault that runs through California. The 1,300 kilometer (800 mile) long fault is the transform boundary between the northwestward-moving Pacific plate and the southeastward-moving North American plate. The San Andreas Fault is a right-lateral strike-slip fault.

The largest earthquake on the San Andreas Fault in historic times occurred in 1906 in San Francisco.

2.2. Earthquakes

An **earthquake** is sudden ground movement caused by the sudden release of energy stored in rocks. The earthquake happens when so much stress builds up in the rocks that the rocks rupture. An earthquake's energy is transmitted by seismic waves.

- During an earthquake, the ground shakes as stored up energy is released from rocks. Nearly all earthquakes occur at plate tectonic boundaries and all types of plate boundaries have earthquakes.
- The Pacific Ocean basin and the Mediterranean-Asiatic belt are the two geographic regions most likely to experience quakes. The seismic waves that do the most damage are surface waves, which only travel along the surface of the ground.
- Body waves travel through the planet and arrive at seismograms before surface waves. Tsunamis are deadly ocean waves that can be caused by undersea earthquakes.

Almost all earthquakes occur at plate boundaries. All three boundary types—divergent, convergent and transform—are prone to earthquake activity. Plate tectonics causes the lithospheric plates to move. When stresses build, they first cause the rocks to bend elastically. If the stresses persist, energy continues to build in the rocks. When the stresses are greater than the internal strength of the rocks, the rocks snap. Although they return to their original shape, the stresses cause the rocks to move to a new position. This movement releases the energy that was stored in the rocks, which creates an earthquake. During an earthquake the rocks usually move several centimeters or maybe as much as a few meters. This description of how earthquakes occur is called **elastic rebound theory** as shown in the Figure below:

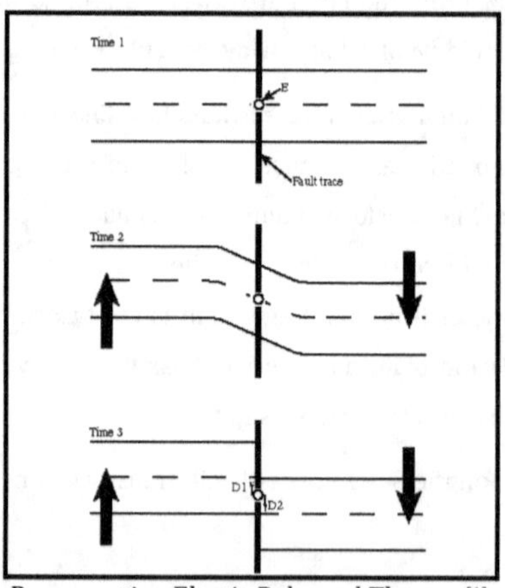

Representing Elastic Rebound Theory. [1]

According to elastic rebound theory, stresses build on both sides of a fault, causing the rocks to deform plastically (Time 2). When the stresses become high enough, the rocks return to their original shape but they move (Time 3). This motion releases the energy that causes an earthquake.

A major **earthquake** 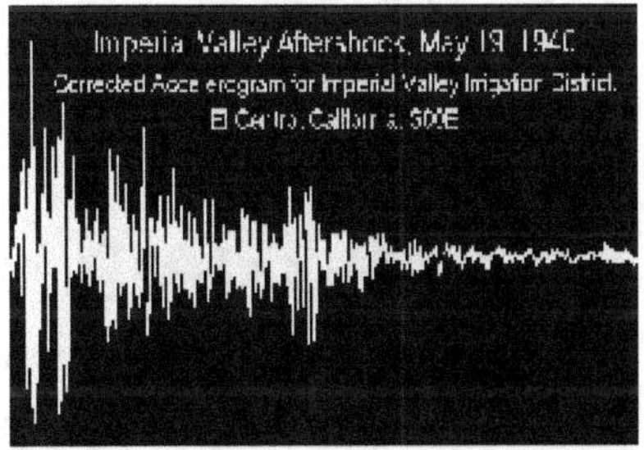 is usually rather **short in duration**, often **lasting only a few seconds** and seldom more than a minute or so.

In general, during a quake there are usually one or more major peaks of magnitude of motion. These peaks represent the maximum effect of the quake.

Although the intensity of the quake is measured in terms of the energy release at the location of the ground fault, the critical effect on the given structure is determined by the ground movements at the location of the structure. The effect of these movements is affected mostly by the distance of the structure from the epicenter, but they are also influenced by the geological conditions directly beneath the structure and by the nature of the entire earth mass between the epicenter and the structure.

Modern recording equipment and practices provide us with representations of the ground movements at various locations, thus allowing us to simulate the **effects of major earthquakes**. One of the most common earthquake representations is acceleration of the ground in one horizontal direction plotted as a function of elapsed time; a typical acceleration record of an earthquake is shown on the Figure below:

Image: Acceleration vs. time record of the 1940 Imperial Valley, California earthquake at the El Centro Station (this is one of the most popular earthquake records)

For use in physical tests in laboratories or in computer modeling, records of actual quakes may be "played back" on structures in order to analyze their responses.

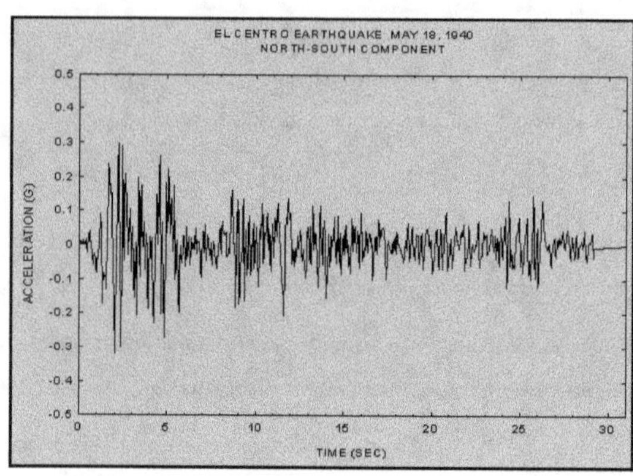

Shaking motion of an earthquake is the result of a sudden release of energy. An earthquake occurs when stress, building up within rocks of the earth's crust, is released in a sudden jolt. Rocks crack and slip past each other causing the ground to vibrate. The slippage emits large amounts of energy in the form of waves that travel through the interior of the earth and across the surface, similar to the waves emanating from a stone dropped into a still pond. An excellent animated story on earthquake waves has been developed by PBS. [2]

Cracks along which rocks slip are called *faults*; these may break through the ground surface, or remain deep within the earth. San Andreas Fault which stretches for more than 900 km along the coast of California is shown on the Figure below:

This fault lies along the boundary between the Pacific Plate and the North American tectonic plate (tectonic plates are discussed in the next section). Activities of this fault have caused some of the major earthquakes in the United States and the world, such as the 1906 San Francisco earthquake (magnitude 8.3 on the Richter scale, killed 700 people and left 250,000 people homeless).

Depending on the relative layout of the rock blocks that move relative to each other in an earthquake and the direction of their relative movements, there are several types of faults. Some of the most common types of faults are: normal fault, reverse fault and strike-slip fault.

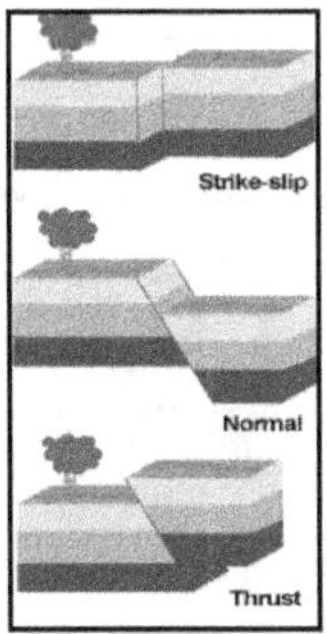

Earthquake Fault Types

Normal Fault: In a normal fault, the block above the fault moves down relative to the block below the fault. This fault motion is caused by tensional forces and results in extension.

Reverse Fault: In a reverse fault, the block above the fault moves up relative to the block below the fault. This fault motion is caused by compression forces and results in shortening. A reverse fault is called ***thrust fault*** if the dip of the fault plane is small.

Strike-Slip Fault: In a strike-slip fault, the movement of blocks along a fault is horizontal. The fault motion of a strike-slip fault is caused by shearing forces.

The location on a fault where slip first occurs is called the ***focus***, whereas the position directly above it on the ground is called the ***epicenter***. Focus of an earthquake and the types of propagating seismic waves (surface waves, "S" waves and "P" waves) are illustrated on the Figure below:

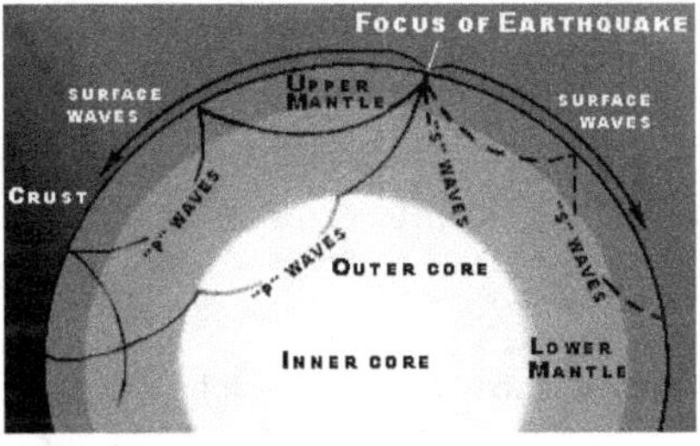

Seismic Waves: Moving and Shaking During an Earthquake

(Source: http://www.windows2universe.org/earth/geology/images/seiamic_waves.gif)

Focal depth is the distance between the focus and the epicenter, whereas distances of a site from the epicenter and focus are known as ***epicentral*** and ***hypocentral distances***, respectively.

Stress that causes an earthquake is created by a movement of almost rigid plates, called **tectonic plates**, which fit together and make up the outer shell of the Earth (also called Earth's crust). These plates float on a dense, liquid layer beneath them. The plates move at such a slow rate (approximately the same rate as a fingernail grows) that the motion is not perceptible. For instance, Juan de Fuca plate off the coast of British Columbia moves only up to 5 cm/year relative to North American plate. Over time, however, this small movement can build up enough stress to produce significant earthquakes. A map illustrating the various tectonic plates that constitute the surface of the earth is shown on the Figure below:

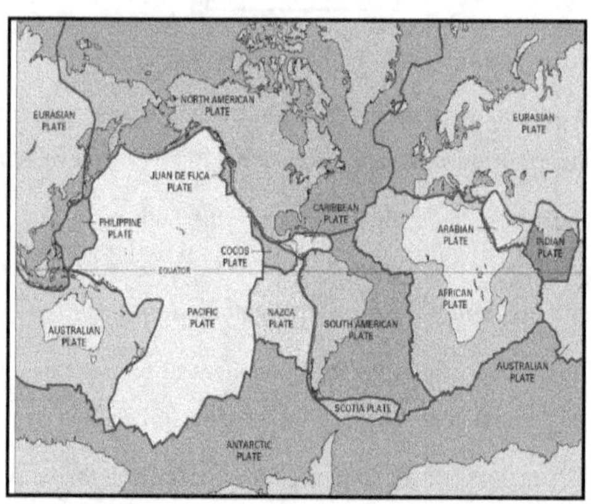

It is to be noted that the western coast of Canada is located at the boundary between two large plates, i.e. the Pacific Plate and the North American Plate (and a small plate called Juan de Fuca plate).

Earthquakes occur most frequently on or near the edges of the plates where stress is most concentrated; such earthquakes are called **interplate** earthquakes. However, a significant number of earthquakes, including some large and damaging ones, do occur within the plates; these earthquakes are known as **intraplate** earthquakes.

2.2.1. *Where Earthquakes Occur*

Earthquakes can strike any location at any time. But history shows they occur in the same general patterns year after year, principally in three large zones of the earth.

The world's greatest earthquake belt, the circum-Pacific seismic belt (also called "The Ring of Fire"), is found along the rim of the Pacific Ocean, where about 81 percent of the world's largest earthquakes occur.

The second important belt, the Alpide, extends from Java to Sumatra through the Himalayas, the Mediterranean, and out into the Atlantic. This belt accounts for about 17 percent of the world's largest earthquakes.

Several major earthquakes that happened in the Washington - B.C. region in the past 150 years are listed in the Table below:

Large Earthquakes in the Past

Year	Region	Magnitude	Comment
1872	Washington-B.C. border	7.4	Widely felt in B.C.
1899	Yukon-Alaska border	8.0	Widely felt in north-western B.C.
1918	Vancouver Island	7.0	Widely felt, minor damage
1929	Queen Charlotte Islands	7.0	Widely felt, minor damage
1946	Vancouver Island	7.3	Widely felt, most damaging quake in western Canada
1949	Queen Charlotte Islands	8.1	**Largest quake in Canada,** one of the world's greatest quakes
1949	Washington	7.0	Much damage in Washington and felt in southwestern B.C.
1958	Alaska-B.C. border	7.9	Damage in Alaska, widely felt in northwestern B.C.
1964	Alaska	9.2	Tsunami damage on Vancouver island
1970	Queen Charlotte Islands	7.4	Widely felt.

2.2.2. *Measuring Earthquakes*

Magnitude and Intensity: The most widely accepted indicators of the size of an earthquake are its magnitude and intensity. The magnitude is a measure of an earthquake in terms of the released energy. At the present time, the most popular scale is the Richter scale, developed by a U.S. seismologist Charles Richter in 1935. Richter defined the magnitude of a local earthquake as the logarithm to base ten of the maximum seismic wave amplitude (in microns) recorded on a standard seismograph at a distance of 100 kilometers from the earthquake epicenter (seismograph is an instrument for recording the motions of the Earth's surface caused by the seismic waves as a function of time). Richter's scale has been recognized by general public, scientists, engineers and technicians as a measure of the relative size of an earthquake. Based on the measurements of seismic energy, it is estimated that each year the total energy released by earthquakes throughout the world is on the order of 1025 to 1026 erg. For the sake of comparison, an atomic bomb blast (e.g. Bikini, 1946) released energy on the order of 1019 erg, whereas an earthquake measuring 5.5 on the Richter scale might release energy of approximately 1020 erg

2.2.3. *Earthquake Magnitude Classes*

Depending on their magnitude, earthquakes are classified into categories ranging from minor to great. Earthquake magnitude classes are shown in the table below:

Class	Magnitude
Great	8 or more
Major	7 - 7.9
Strong	6 - 6.9
Moderate	5 - 5.9
Light	4 - 4.9
Minor	3 -3.9

The Table below briefly describes earthquake effects corresponding to various magnitude levels and also gives an estimated number of earthquakes of different magnitudes that happen in the world each year. It can be observed from this table that a large majority of earthquakes (900,000) are of magnitude 2.5 or less (very minor earthquakes, usually not felt). Great, catastrophic earthquakes (magnitude 8 or greater) happen once in 5 to 10 years.

Earthquake Magnitude Scale

Magnitude	Earthquake Effects	Estimated Number Each Year
2.5 or less	Usually not felt, but can be recorded by seismograph.	900,000
2.5 to 5.4	Often felt, but only causes minor damage.	30,000
5.5 to 6.0	Slight damage to buildings and other structures.	500
6.1 to 6.9	May cause a lot of damage in very populated areas.	100
7.0 to 7.9	Major earthquake. Serious damage.	20
8.0 or greater	Great earthquake. Can totally destroy communities near the epicenter.	One every 5 to 10 years

The assessment of earthquake intensity on a descriptive scale depends on actual observations of earthquake effects. Observations on the performance of building structures, natural phenomena, and human perceptions are essential for evaluating the earthquake intensity. Intensity of an earthquake depends on the distance from epicenter, and also on the local soil conditions, geology and topography. In a typical case, however, the largest intensity is observed in the vicinity of epicenter and it diminishes with the distance.

The intensity scale consists of a series of certain key responses such as people awakening, movement of furniture, damage to chimneys, and finally-total destruction. Although numerous intensity scales have been developed over the last several hundred years to evaluate the effects of earthquakes, the one currently used in the United States is the ***Modified Mercalli Intensity (MMI) Scale***. This scale, composed of 12 increasing levels of intensity that range from imperceptible shaking to catastrophic destruction, is designated by Roman numerals. It does not have a mathematical basis; instead it is an arbitrary ranking based on observed effects.

The lower numbers of the intensity scale generally deal with the manner in which the earthquake is felt by people. The higher numbers of the scale are based on observed structural damage. Structural engineers usually contribute information for assigning intensity values of VIII or above.

Detailed specifications of various intensity levels related to the MMI intensity scale are presented in the Table below:

Modified Mercalli Intensity (MMI) Scale

MMI Level	Description
I	Not felt except by a very few under favourable circumstances. Typically marginal and long-period effects of strong earthquakes.
II	Felt only by a few persons at rest, especially on the upper floors of buildings. Delicately suspended objects may swing.
III	Felt quite noticeably indoors, especially on upper floors of buildings, but many people do not recognize it as an earthquake. Standing cars may rock slightly. Vibration similar to that of passing truck. Duration may be estimated.
IV	If during the day, felt indoors by many; outdoors by few. If at night, few awakened. Dishes, windows and doors rattle, walls creak. A sensation such as heavy truck shaking the building. Standing cars rock noticeably. In the upper range of IV wooden walls and frames creak.
V	Felt by nearly everyone, many awakened; direction may be estimated. Some dishes and windows broken, some plaster cracked, unstable objects overturned. Disturbance of trees, poles and other tall objects. Pendulum clocks may stop.
VI	Felt by all; many people frightened and run outdoors. Persons walk unsteadily. Windows, dishes, glassware broken. Books off shelves; pictures off walls. Weak plaster and masonry D cracked; minor chimney damage. Movement of moderately heavy furniture; some furniture overturned. Small bells ring (school, church). Trees, bushes shaken visibly, or heard to rustle.
VII	Everybody runs outdoors. Noticed by persons driving cars. Hanging objects quiver. Furniture broken. Damage to masonry D3, including cracks. Some cracks in masonry C. Weak chimneys broken at roof line. Fall of plaster, loose bricks, stones, tiles, cornices, unbraced parapets and architectural ornaments. Waves on pond; water turbid with mud. Small slides and caving in along sand or gravel banks. Large bells ring. Concrete irrigation ditches damaged.
VIII	Persons driving in cars are disturbed. Damage to masonry C; partial collapse. Some damage to masonry B; none to masonry A. Fall of stucco and some masonry walls. Panel walls thrown out of frame structures. Chimneys, factory stacks, monuments, walls, and columns fall. Heavy furniture overturned or damaged. Changes in well water. Sand and mud ejected in small amounts. Cracks in wet ground and on steep slopes. Branches broken from the trees.
IX	General panic. Masonry D destroyed; masonry C heavily damaged, sometimes with complete collapse; masonry B seriously damaged. General damage to foundations. Frame structures, if not bolted, shifted off foundations. Frames racked. Serious damage to reservoirs. Ground noticeably cracked, underground pipes broken. In alluviated areas sand and mud ejected, earthquake fountains and sand craters appear.
X	Most masonry and frame structures destroyed with their foundations. Some well-built wooden structures destroyed, foundations ruined, ground badly cracked. Serious damage to dams, dikes, embankments. Rails bent slightly. Considerable landslides from steep slopes and river banks. Water splashed over banks. Sand and mud shifted horizontally on beaches and flat land.
XI	Rails bent greatly. Bridges destroyed. Broad fissures in ground. Underground pipes out of service. Earth slumps and land slips in soft ground.
XII	Total damage. Waves are seen on the ground surface. Large rock masses displaced. Lines of sight and level are distorted. Objects thrown into the air.

A major difference between the earthquake intensity and magnitude lies in the fact that magnitude of an earthquake is determined based on measuring the ground motion with instruments (seismographs), whereas the intensity of an earthquake is determined based on observations of earthquake effects on building structures and human perceptions.

Another essential difference between a magnitude and intensity of an earthquake lies in the fact that magnitude is a unique indicator of a size of an earthquake - each earthquake is characterized with a single value which indicates its magnitude. At the same time, each earthquake is characterized with various intensities, depending on the location of a particular site with respect to the epicenter. For example, Canada's largest historic earthquake, the Queen Charlotte Island earthquake of August 22, 1949 was characterized with magnitude 8.1 on the Richter scale. The same earthquake was characterized with MMI intensities ranging from III to over VII, as illustrated in the figure below.

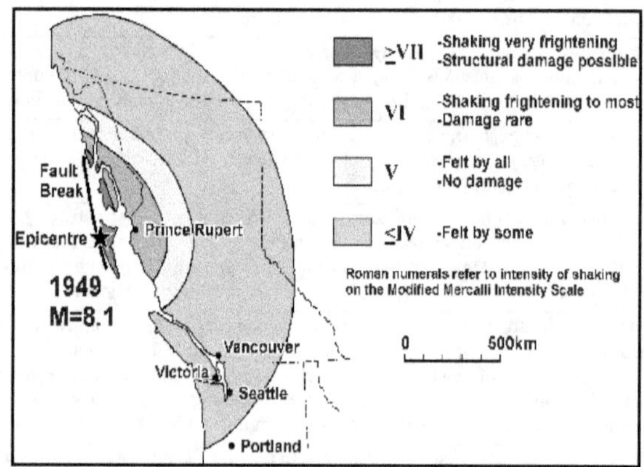

As an illustration of MMI intensity of VII or higher in the area close to the epicenter of this earthquake "cows were knocked off their feet, and a geologist with the Geological Survey of Canada working on the north end of Graham Island could not stand up." In Prince Rupert (MMI intensity VI), "windows were shattered and buildings swayed."

References

1. https://en.wikibooks.org/wiki/High_School_Earth_Science/Nature_of_Earthquakes
2. http://www.pbs.org/wnet/savageearth/animations/earthquakes/main.html

3. Inner Structure of the Earth

Abstract

A study of the earth's interior structure is quite helpful in understanding the science behind the earthquake. At the present level of understandings, earthquakes are usually caused when rock underground suddenly breaks along a fault. This sudden release of energy causes the seismic waves that make the ground shake. When two blocks of rock or two plates are rubbing against each other, they stick a little. They don't just slide smoothly; the rocks catch on each other. The rocks are still pushing against each other, but not moving. After a while, the rocks break because of all the pressure that's built up. When the rocks break, the earthquake occurs. During the earthquake and afterward, the plates or blocks of rock start moving, and they continue to move until they get stuck again. The spot underground where the rock breaks is called the **focus** of the earthquake. The place right above the focus (on top of the ground) is called the **epicenter** of the earthquake. Movement of seismic waves from the point of focus up to the earth's surface through different layers of rocks/plates governs the magnitude and severity of the earthquake.

The structure of the Earth: The Earth consists of four concentric layers: inner core, outer core, mantle and crust. The crust is made up of tectonic plates, which are in constant motion. Earthquakes and volcanoes are most likely to occur at plate boundaries.The four distinct layers constituting the earth are as given below:

1. **The inner core** is in the centre and is the hottest part of the Earth. It is solid and made up of iron and nickel with temperatures of up to 5,500°C. With its immense heat energy, the inner core is like the engine room of the Earth.

2. **The outer core** is the layer surrounding the inner core. It is a liquid layer, also made up of iron and nickel. It is still extremely hot, with temperatures similar to the inner core.

3. **The mantle** is the widest section of the Earth. It has a thickness of approximately 2,900 km. The mantle is made up of semi-molten rock called magma. In the upper parts of the mantle the rock is hard, but lower down the rock is soft and beginning to melt.

4. **The crust** is the outer layer of the earth. It is a thin layer between 0-60 km thick. The crust is the solid rock layer upon which we live.

It is important to mention that there are two different types of crust: **continental crust**, carrying land, and **oceanic crust**, which carries water.

The diagram below shows the structure of the earth. In geography, taking a slice through a structure to see inside is called a **cross section**.

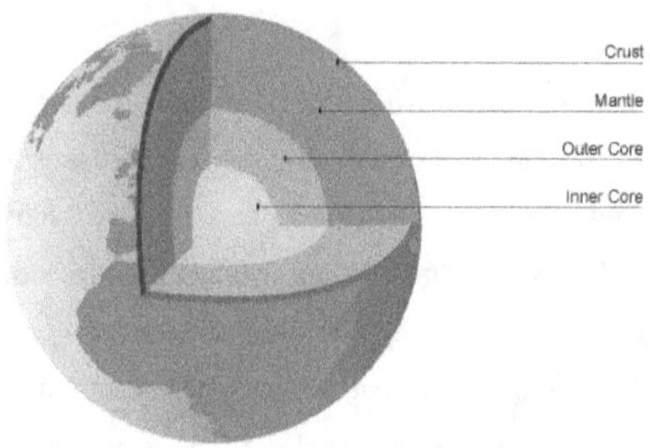

Cross section showing structure of the Earth

(Source: http://www.bbc.co.uk/staticarchive/18e0518ed5183cd531f908f11031e176c4461b9c.gif)

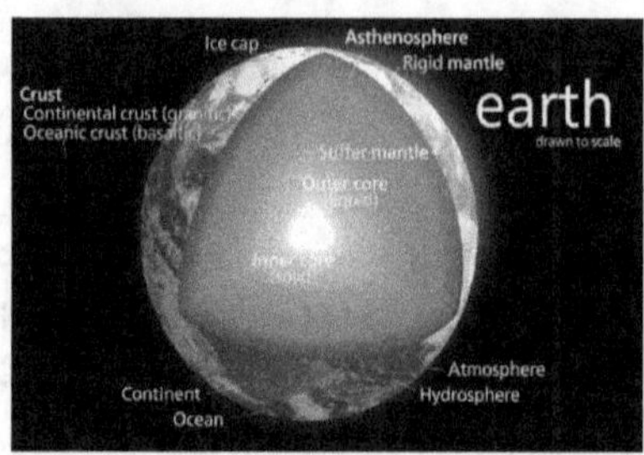

Structure of the Earth

3.1.   Earth's Interior

The interior **structure of the Earth** is layered in spherical shells. These layers can be defined by either their chemical or their rheological properties. Earth has an outer silicate solid crust, a highly viscous mantle, a liquid outer core that is much less viscous than the mantle, and a solid inner core. Scientific understanding of the internal structure of the Earth is based on observations of topography and bathymetry, observations of rock in outcrop, samples brought to the surface from greater depths by volcanic activity, analysis of the seismic waves that pass through the Earth, measurements of the gravitational and magnetic fields of the Earth, and experiments with crystalline solids at pressures and temperatures characteristic of the Earth's deep interior.

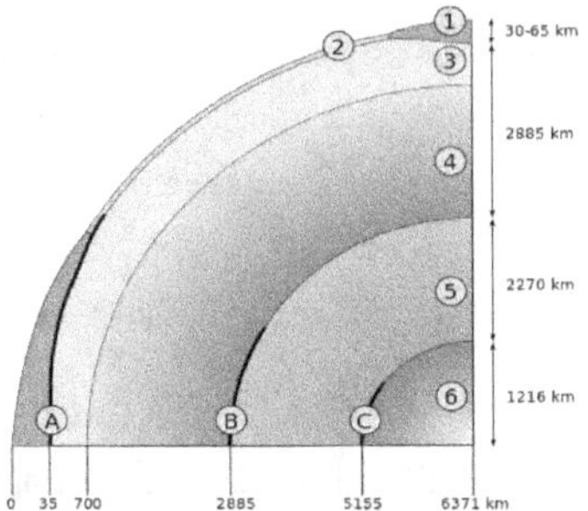

Schematic representation: Inner Structure of the Earth

1. Continental crust

2. Oceanic crust

3. Upper mantle

4. Lower mantle

5. Outer core,

6. Inner core

A: Mohorovičić discontinuity

B: Gutenberg discontinuity

C: Lehmann–Bullen discontinuity

The structure of Earth can be defined either by mechanical properties such as rheology or by chemical properties.

Mechanically, it can be divided into lithosphere, asthenosphere, mesospheric mantle, outer core, and the inner core. The interior of Earth is divided into 5 important layers. Chemically, Earth can be divided into the crust, upper mantle, lower mantle, outer core, and inner core. The geologic component layers of Earth [1] are at the following depths below the surface:

Important layers of the earth

Depth		Layer
Kilometes	Miles	
0–60	0–37	Lithosphere (locally varies between 5 and 200 km)
0–35	0–22	Crust (locally varies between 5 and 70 km)
35–60	22–37	Uppermost part of mantle
35–2,890	22–1,790	Mantle
100–200	210-270	Upper mesosphere (upper mantle)
660–2,890	410–1,790	Lower mesosphere (lower mantle)
2,890–5,150	1,790–3,160	Outer core
5,150–6,360	3,160–3,954	Inner core

The layering of Earth has been inferred indirectly using the time of travel of refracted and reflected seismic waves created by earthquakes. The core does not allow shear waves to pass through it, while the speed of travel (seismic velocity) is different in other layers. The changes in seismic velocity between different layers causes refraction owing to Snell's law, like light bending as it passes through a prism. Likewise, reflections are caused by a large increase in seismic velocity and are similar to light reflecting from a mirror.

3.1.1. Core: Inner Core and Outer Core

The average density of Earth is 5,515 kg/m³. Since the average density of surface material is only around 3,000 kg/m³, we must conclude that denser materials exist within Earth's core. *Seismic measurements show that the core is divided into two parts, a "solid" inner core with a radius of ~1,220 km*[2] and a liquid outer core extending beyond it to a radius of ~3,400 km. The densities are between 9,900 and 12,200 kg/m³ in the outer core and 12,600–13,000 kg/m³ in the inner core.[3]

3.1.2. Mantle: Mantle (Geology)

Earth's mantle extends to a depth of 2,890 km, making it the thickest layer of Earth. The upper mantle is divided into the lithospheric mantle and the asthenosphere. The upper and lower mantles are separated by the transition zone. The lowest part of the mantle next to the core-mantle boundary is known as the D″ layer. The pressure at the bottom of the mantle is ~140 GPa (1.4 Matm). The mantle is composed of silicate rocks that are rich in iron and magnesium relative to the overlying crust. Although solid, the high temperatures within the mantle cause the silicate material to be sufficiently ductile that it can flow on very long timescales. Convection of the mantle is expressed at the surface through the motions of tectonic plates. As there is intense and increasing pressure as one travels deeper into the mantle, the lower part of the mantle flows less easily than does the upper mantle (chemical changes within the mantle may also be important). The viscosity of the mantle ranges between 10^{21} and 10^{24} Pa·s, depending on depth.[4] In comparison, the viscosity of water is approximately 10^{-3} Pa·s and that of pitch is 10^7 Pa·s.

3.1.3. Crust: Crust (geology)

The crust ranges from 5–70 km (~3–44 miles) in depth and is the outermost layer. The thin parts are the oceanic crust, which underlie the ocean basins (5–10 km) and are composed of dense(mafic)iron magnesium silicate igneous rocks, like basalt. The thicker crust is continental crust, which is less dense and composed of (felsic) sodium potassium aluminium silicate rocks, like granite. The rocks of the crust fall into two major categories – sial and sima (Suess,1831–

1914). It is estimated that sima starts about 11 km below the Conrad discontinuity (a second order discontinuity).The uppermost mantle together with the crust constitutes the lithosphere. The crust-mantle boundary occurs as two physically different events. First, there is a discontinuity in the seismic velocity, which is known as the Mohorovičić discontinuity or Moho. The cause of the Moho is thought to be a change in rock composition from rocks containing plagioclase feldspar(above) to rocks that contain no feldspars(below). Second, in oceanic crust, there is a chemical discontinuity between ultramafic cumulates and tectonized harzburgites, which has been observed from deep parts of the oceanic crust that have been obducted onto the continental crust and preserved as ophiolite sequences.

Many rocks now making up Earth's crust formed less than 100 million (1×10^8) years ago; however, the oldest known mineral grains are 4.4 billion (4.4×10^9) years old, indicating that Earth has had a solid crust for at least that long.[5]

3.2.   The Structure of the Earth: Tectonic Plates [6]

The Earth consists of four concentric layers: inner core, outer core, *mantle* and *crust*. The crust is made up of *tectonic* plates, which are in constant motion. Earthquakes and volcanoes are most likely to occur at plate boundaries.

The four distinct concentric layers are:

- **The inner core** is in the centre and is the hottest part of the Earth. It is solid and made up of iron and nickel with temperatures of up to 5,500°C. With its immense heat energy, the inner core is like the engine room of the Earth.
- **The outer core** is the layer surrounding the inner core. It is a liquid layer, also made up of iron and nickel. It is still extremely hot, with temperatures similar to the inner core.
- **The mantle** is the widest section of the Earth. It has a thickness of approximately 2,900 km. The mantle is made up of semi-molten rock called magma. In the upper parts of the mantle the rock is hard, but lower down the rock is soft and beginning to melt.
- **The crust** is the outer layer of the earth. It is a thin layer between 0-60 km thick. The crust is the solid rock layer upon which we all live.

There are two different types of crust: **continental crust**, which carries land, and **oceanic crust**, which carries water.

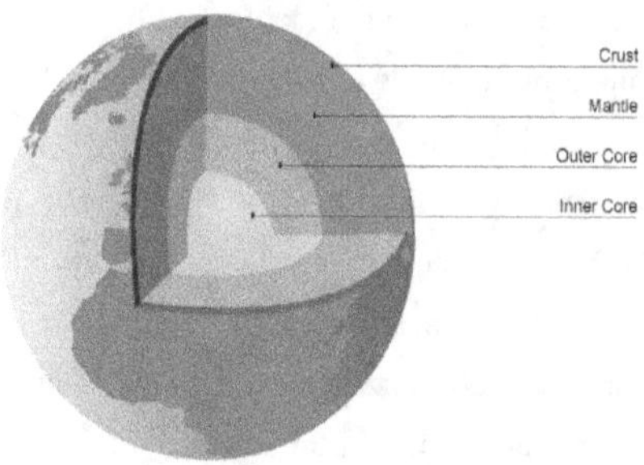

Four Distinct Layers of the earth's Structure

3.2.1. *Distribution: Tectonic Plates* [7]

The Earth's crust is broken up into pieces called plates. Heat rising and falling inside the mantle creates **convection currents** generated by radioactive decay in the core. The convection currents move the plates. Where convection currents diverge near the Earth's crust, plates move apart. Where convection currents converge, plates move towards each other. The movement of the plates, and the activity inside the Earth, is called **plate tectonics**.

Plate tectonics cause earthquakes and volcanoes. The point where two plates meet is called a **plate boundary**. Earthquakes and volcanoes are most likely to occur either on or near plate boundaries.

It is to be mentioned that the Earth's plates move in different directions.

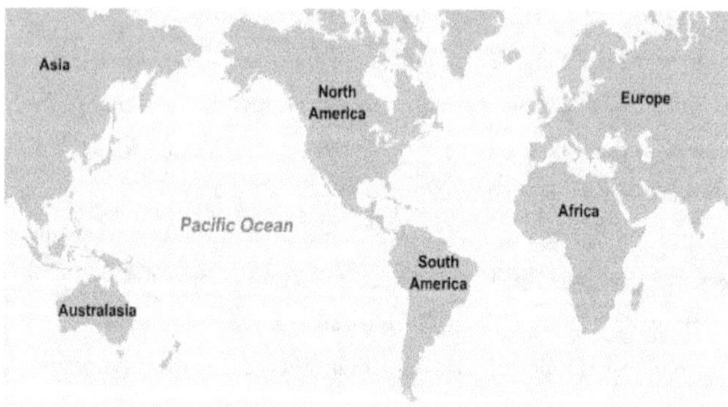

The map shows the world's tectonic plates and the distribution of earthquakes and volcanoes

3.2.2. Different Plate Boundaries

- At a **tensional, constructive** or **divergent boundary** the plates move apart.
- At a **compression, destructive** or **convergent** boundary the plates move towards each other.
- At a **conservative** or **transform** boundary the plates slide past each other.
- **Tensional Margins** [8]

At a tensional or **constructive boundary** the plates are moving apart. The plates move apart due to *convection currents* inside the Earth.

The Helgafjell Volcano on Westman Island, Iceland

As the plates move apart (very slowly), *magma* rises from the mantle. The magma erupts to the surface of the Earth. This is also accompanied by earthquakes.

When the magma reaches the surface, it cools and solidifies to form a new crust of **igneous rock**. This process is repeated many times, over a long period of time.

Eventually the new rock builds up to form a volcano. **Constructive boundaries** tend to be found under the sea, eg the Mid Atlantic Ridge. Here, chains of underwater volcanoes have formed along the *plate boundary*. One of these volcanoes may become so large that it erupts out of the sea to form a volcanic island, e.g., Surtsey and the Westman Islands near Iceland. The animation of diagram below shows how magma pushes up between the two plates, causing a chain of volcanoes along the constructive plate boundary.

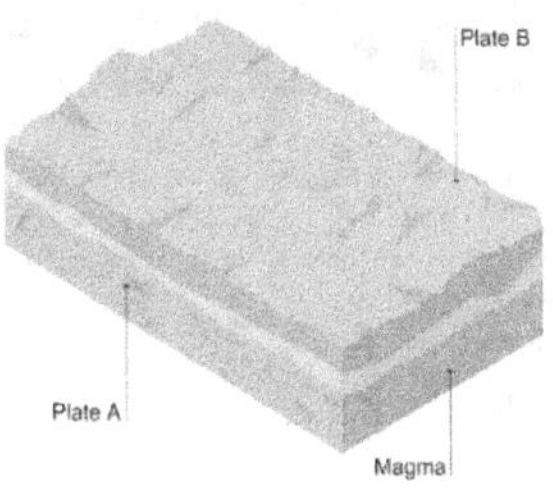

- ### *Compression Boundary* [9]

At a compression or **destructive boundary,** the plates are moving towards each other. This usually involves a **continental plate** and an **oceanic plate.**The oceanic plate is **denser** than the continental plate so, as they move together, the oceanic plate is forced underneath the continental plate. The point at which this happens is called the **subduction zone**. As the oceanic plate is forced below the continental plate it melts to form magma and earthquakes are triggered. The magma collects to form a *magma chamber*. This magma then rises up through cracks in the continental crust. As pressure builds up, a volcanic eruption may occur.

The animation of diagram below shows how the oceanic plate is pushed underneath the continental plate, causing mountains and possibly volcanoes to form along the destructive plate boundary.

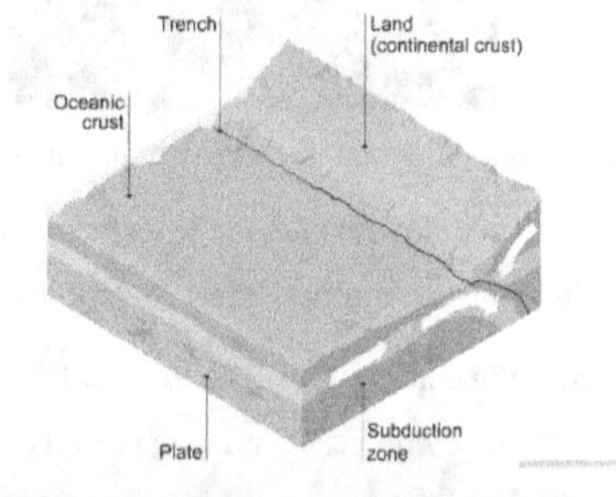

Image: A View of the Himalayas from Gorak Shep

As the plates push together, the continental crust is squashed together and forced upwards. This is called folding. The process of folding creates *fold mountains*. Fold mountains can also be formed where two continental plates push towards each other. This is how mountain ranges such as the **Himalayas** and the **Alps** were formed.

3.2.3. *Earth's Interior & Plate Tectonics* [10]

We must pay attention and listen to the ring and vibration of our Earth in an attempt to discover its content. This is accomplished through seismology, which has become the principle method used in studying Earth's interior. Seismology on Earth deals with the study of vibrations that are produced by earthquakes. A seismograph is used to measure and record the actual movements and vibrations within the Earth and of the ground.

Scientists categorize seismic movements into four types of diagnostic waves that travel at speeds ranging from 3 to 15 kilometers (1.9 to 9.4 miles) per second. Two of the waves travel around the surface of the Earth in rolling swells. The other two, Primary (P) or compression waves and Secondary (S) or shear waves, penetrate the interior of the Earth. Primary waves compress and dilate the matter they travel through (either rock or liquid) similar to sound waves. They also have the ability to move twice as fast as S waves. Secondary waves propagate through rock but are not able to travel through liquid. Both P and S waves refract or reflect at points where layers of differing physical properties meet. They also reduce speed when moving through hotter material. These changes in direction and velocity are the means of locating discontinuities.

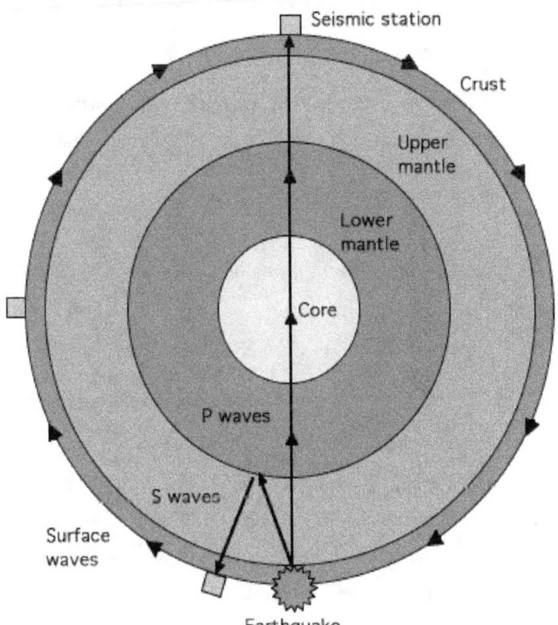

Types of Seismic Waves (Adapted from, Beatty, 1990.)

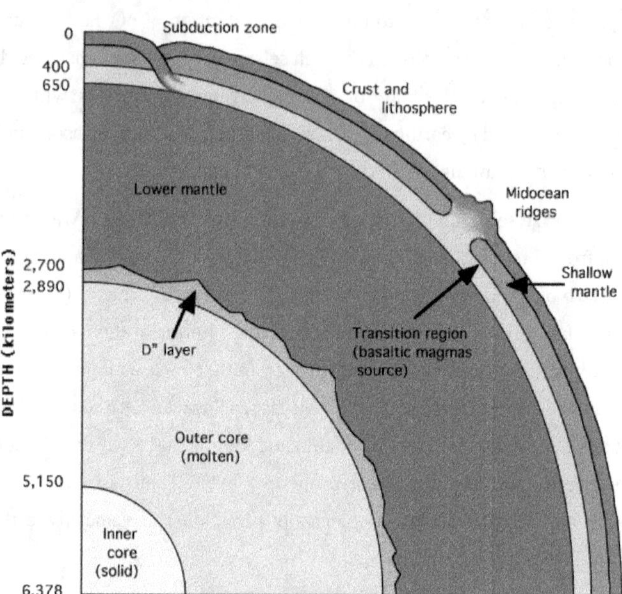

3.2.4. *Different Layers in the Earth's Interior* [11]

Seismic discontinuities aid in distinguishing divisions of the Earth into inner core, outer core, D" layer, lower mantle, transition region, upper mantle, and crust (oceanic and continental).

Inner core: 1.7% of the Earth's mass; depth of 5,150-6,370 kilometers (3,219 - 3,981 miles)

The inner core is solid and unattached to the mantle, suspended in the molten outer core. It is believed to have solidified as a result of pressure-freezing which occurs to most liquids when temperature decreases or pressure increases.

- **Outer core: 30.8% of Earth's mass; depth of 2,890-5,150 kilometers (1,806 - 3,219 miles)**

The outer core is a hot, electrically conducting liquid within which convective motion occurs. This conductive layer combines with Earth's rotation to create a dynamo effect that maintains a system of electrical currents known as the Earth's magnetic field. It is also responsible for the subtle jerking of Earth's rotation.

This layer is not as dense as pure molten iron, which indicates the presence of lighter elements. Scientists suspect that about 10% of the layer is composed of sulfur and/or oxygen because these elements are abundant in the cosmos and dissolve readily in molten iron.

- *D" layer: 3% of Earth's mass; depth of 2,700-2,890 kilometers (1,688 - 1,806 miles)*

This layer is 200 to 300 kilometers (125 to 188 miles) thick and represents about 4% of the mantle-crust mass. Although it is often identified as part of the lower mantle, seismic discontinuities suggest the D" layer might differ chemically from the lower mantle lying above it. Scientists theorize that the material either dissolved in the core, or was able to sink through the mantle but not into the core because of its density.

- *Lower mantle: 49.2% of Earth's mass; depth of 650-2,890 kilometers (406 -1,806 miles)*

The lower mantle contains 72.9% of the mantle-crust mass and is probably composed mainly of silicon, magnesium, and oxygen. It probably also contains some iron, calcium, and aluminum. Scientists make these deductions by assuming the Earth has a similar abundance and proportion of cosmic elements as found in the Sun and primitive meteorites.

- *Transition region: 7.5% of Earth's mass; depth of 400-650 kilometers (250-406 miles)*

The transition region or mesosphere (for middle mantle), sometimes called the fertile layer, contains 11.1% of the mantle-crust mass and is the source of basaltic magmas. It also contains calcium, aluminum, and garnet, which is a complex aluminum-bearing silicate mineral. This layer is dense when cold because of the garnet. It is buoyant when hot because these minerals melt easily to form basalt which can then rise through the upper layers as magma.

- *Upper mantle: 10.3% of Earth's mass; depth of 10-400 kilometers (6 - 250 miles)*

The upper mantle contains 15.3% of the mantle-crust mass. Fragments have been excavated for our observation by eroded mountain belts and volcanic eruptions. Olivine $(Mg,Fe)_2SiO_4$ and pyroxene $(Mg,Fe)SiO_3$ have been the primary minerals found in this way. These and other minerals are refractory and crystalline at high temperatures; therefore, most settle out of rising magma, either forming new crustal material or never leaving the mantle. Part of the upper mantle called the asthenosphere might be partially molten.

- ***Oceanic crust: 0.099% of Earth's mass; depth of 0-10 kilometers (0 - 6 miles)***

The oceanic crust contains 0.147% of the mantle-crust mass. The majority of the Earth's crust was made through volcanic activity. The oceanic ridge system, a 40,000-kilometer (25,000 mile) network of volcanoes, generates new oceanic crust at the rate of 17 km³ per year, covering the ocean floor with basalt. Hawaii and Iceland are two examples of the accumulation of basalt piles.

- ***Continental crust: 0.374% of Earth's mass; depth of 0-50 kilometers (0 - 31 miles).***

The continental crust contains 0.554% of the mantle-crust mass. This is the outer part of the Earth composed essentially of crystalline rocks. These are low-density buoyant minerals dominated mostly by quartz (SiO2) and feldspars (metal-poor silicates). The crust (both oceanic and continental) is the surface of the Earth; as such, it is the coldest part of our planet. Because cold rocks deform slowly, we refer to this rigid outer shell as the lithosphere (the rocky or strong layer).

3.2.5. The Lithosphere & Plate Tectonics: [12] [13] [14]

- ***Oceanic Lithosphere***

The rigid, outermost layer of the Earth comprising the crust and upper mantle is called the lithosphere. New oceanic lithosphere forms through volcanism in the form of fissures at mid-ocean ridges which are cracks that encircle the globe. Heat escapes the interior as this new lithosphere emerges from below. It gradually cools, contracts and moves away from the ridge, traveling across the seafloor to subduction zones in a process called seafloor spreading. In time, older lithosphere will thicken and eventually become more dense than the mantle below, causing it to descend (subduct) back into the Earth at a steep angle, cooling the interior. Subduction is the main method of cooling the mantle below 100 kilometers (62.5 miles). If the lithosphere is young and thus hotter at a subduction zone, it will be forced back into the interior at a lesser angle.

- ***Continental Lithosphere***

The continental lithosphere is about 150 kilometers (93 miles) thick with a low-density crust and upper-mantle that are permanently buoyant. Continents drift laterally along the convecting system of the mantle away from hot mantle zones toward cooler ones, a process known as continental drift. Most of the continents are now sitting on or moving toward cooler parts of the mantle, with the exception of Africa.

Africa was once the core of Pangaea, a supercontinent that eventually broke into todays continents. Several hundred million years prior to the formation of Pangaea, the southern continents - Africa, South America, Australia, Antarctica, and India - were assembled together in what is called Gondwana.

- ***Plate Tectonics***

Image: Crustal Plate Boundaries (Courtesy NGDC)

Plate tectonics involves the formation, lateral movement, interaction, and destruction of the lithosphere plates. Much of Earth's internal heat is relieved through this process and many of Earth's large structural and topographic features are consequently formed. Continental rift valleys and vast plateaus of basalt are created at plate break up when magma ascends from the mantle to the ocean floor, forming new crust and separating mid ocean ridges. Plates collide and are destroyed as they descend at subduction zones to produce deep ocean trenches, strings of volcanoes, extensive transformed faults, broad linear rises, and folded mountain belts. Earth's lithosphere presently is divided into eight large plates with about two dozen smaller ones that are drifting above the mantle at the rate of 5 to 10 centimeters (2 to 4 inches) per year. The eight large plates are the African, Antarctic, Eurasian, Indian-Australian, Nazca, North American, Pacific, and South American plates. A few of the smaller plates are the Anatolian, Arabian, Caribbean, Cocos, Philippine, and Somali plates.

3.3. Inner-core and Outer-core [15]

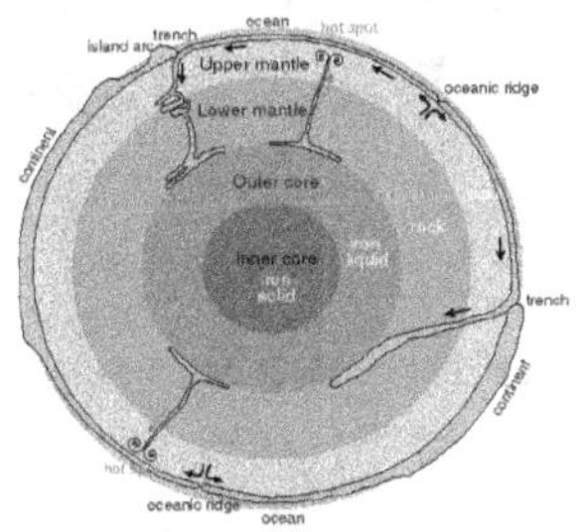

3.3.1. *Inner Core* [16]

The earth's inner core is the earth's innermost part and according to seismological studies, it is primarily a solid ball with a radius of about 1220 kilometers.

3.3.2. *Outer Core* [16]

This is a liquid layer about 2,300 km (1,400 mi)[1] thick and composed of iron and nickel that lies above Earth's solid inner core and below its mantle. Its outer boundary lies 2,890 km (1,800 mi) beneath Earth's surface. The transition between the inner core and outer core is located approximately 5,150 km (3,200 mi) beneath the Earth's surface.

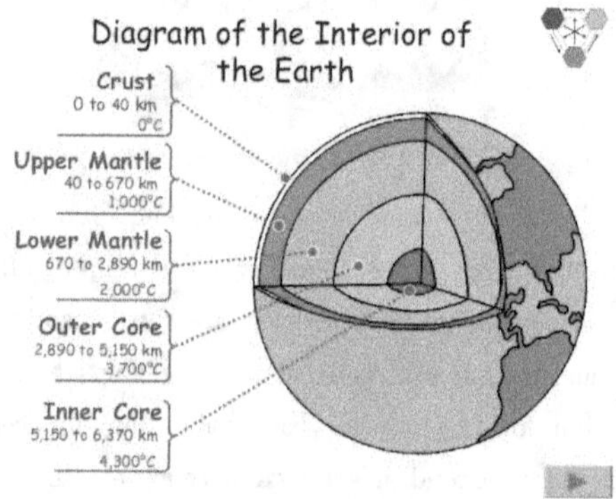

Image: Interior of the Earth

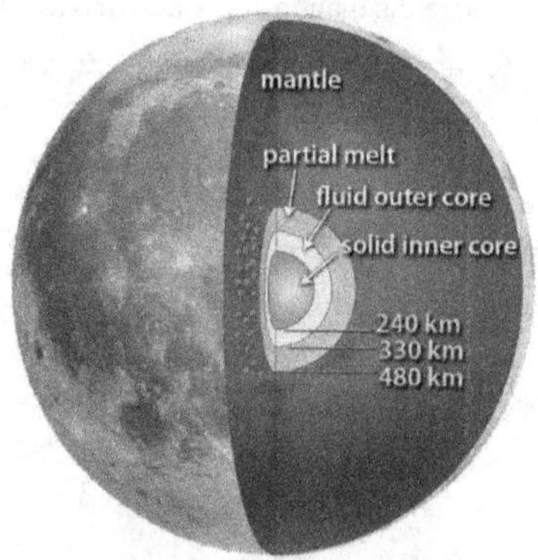

Image: Moon Core Composition | 868 x 866 · 397 kB · jpeg

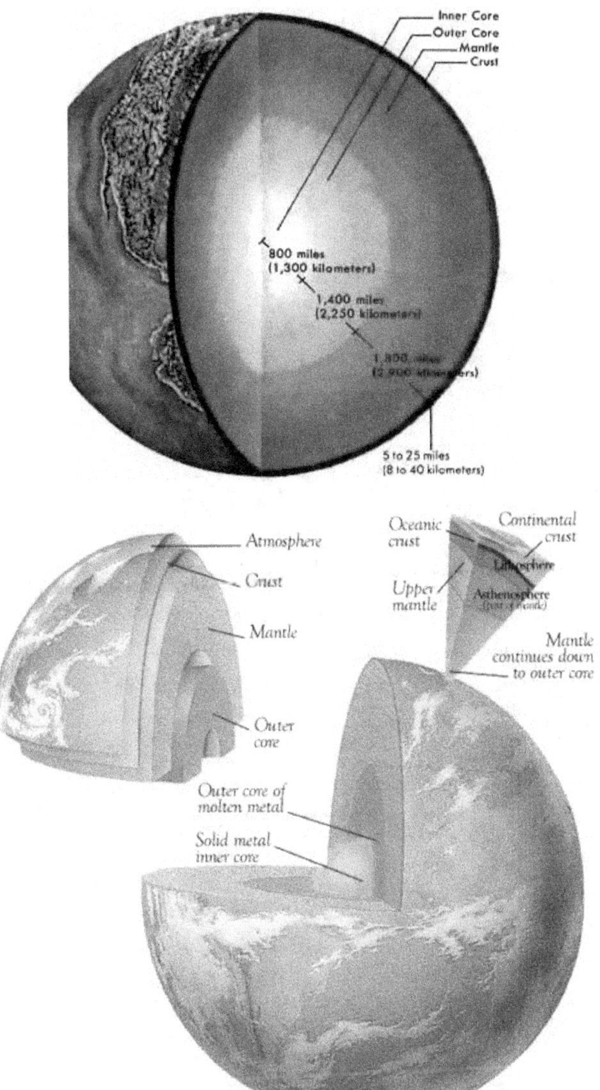
Inner Core
Outer Core
Mantle
Crust
800 miles
(1,300 kilometers)
1,400 miles
(2,250 kilometers)
1,800 miles
(2,900 kilometers)
5 to 25 miles
(8 to 40 kilometers)
Atmosphere
Crust
Mantle
Outer core
Oceanic crust
Continental crust
Lithosphere
Upper mantle
Asthenosphere
(part of mantle)
Mantle continues down to outer core
Outer core of molten metal
Solid metal inner core

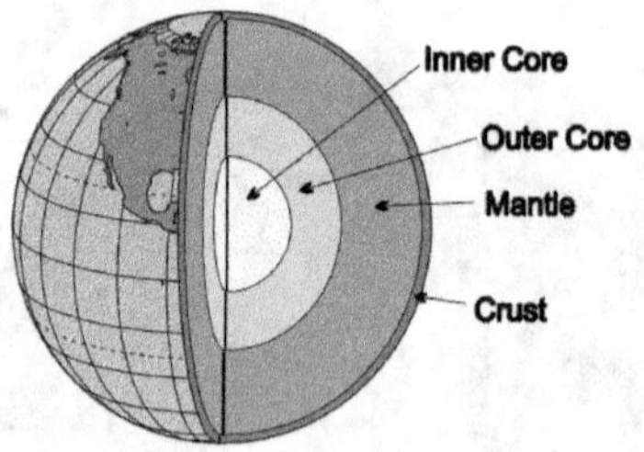

3.3.3.   Third Rock from the Sun - Restless Earth [17]

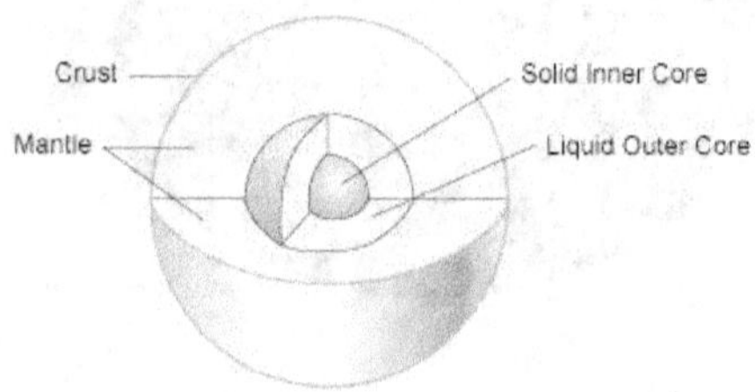

Restless Earth

A relatively thin, rocky crust covers a thick silicate mantle. They overlie a liquid outer core, composed mainly of molten iron, and an inner core of solid iron. These nested layers have been inferred from seismic waves that travel through the Earth, changing velocity and direction at the layer boundaries (Figs. A, B) below:

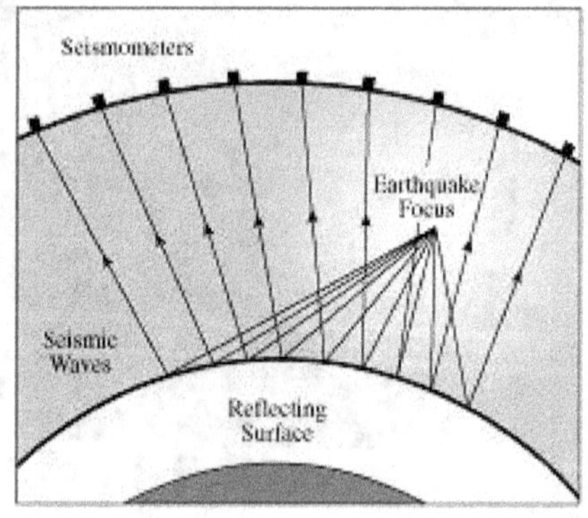

Figure: A

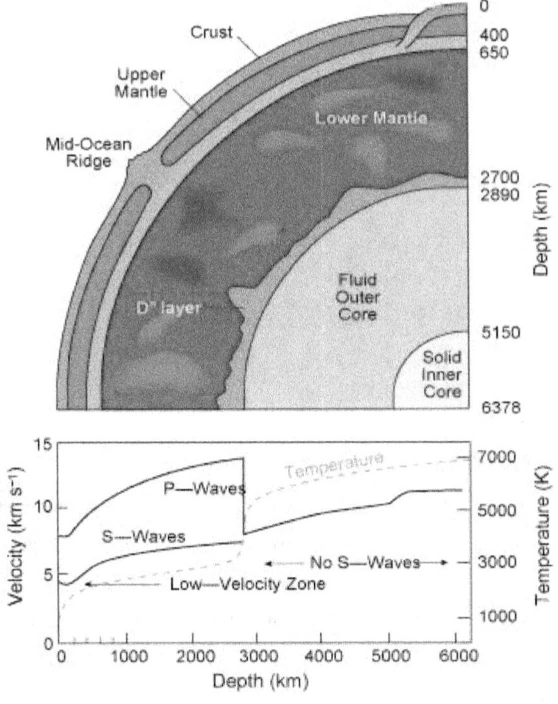

Figure: B

3.3.4. *Layered Structure of the Earth*

The Earth's internal structure is determined by the varying velocity of earthquake wavesas ashown in the above Figure. There are two types of waves that travel through the Earth. They are known as the compression P, or push and pull, waves and the shear S, or shake, waves. The P waves move almost twice as fast as the S waves, and the P waves pass through the fluid outer core which the S waves cannot do. The boundary between the mantle and core is marked be a precipitous drop in the velocity of the P waves at a depth of about 2.9 million meters. The S waves do not propagate beyond this boundary. The liquid outer core is separated from the solid inner core at a radius of 1.22 thousand meters where the P waves increase in velocity.

The Earth's double core

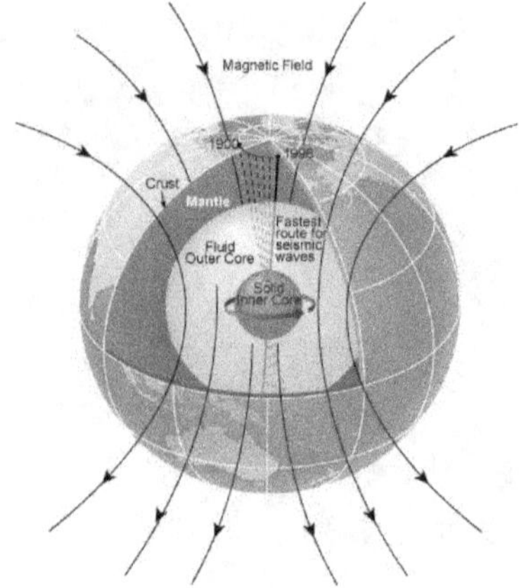

The Earth's Double Core

The mantle and part of the crust have been cut away here to show the relative sizes of the Earth's fluid and solid cores. The outer fluid core is about 55 percent of the radius of the Earth, and the inner solid core is slightly smaller than the Moon. The Earth's magnetic field is thought to be generated and sustained by moving currents in the planet's electrically-conducting, fluid outer core, which is composed of molten iron. Geophysicists have discovered that the route of the rapid polar (north-south) waves through the Earth's interior is gradually shifting eastward because the inner core is rotating slightly faster than the rest of the planet. The fast rotation of the inner solid core may help explain how Earth's magnetic field reverses polarity (Courtesy of Paul Richards, Lamont-Doherty Earth Observatory.)

3.3.5.   *Hot Zone*[18]

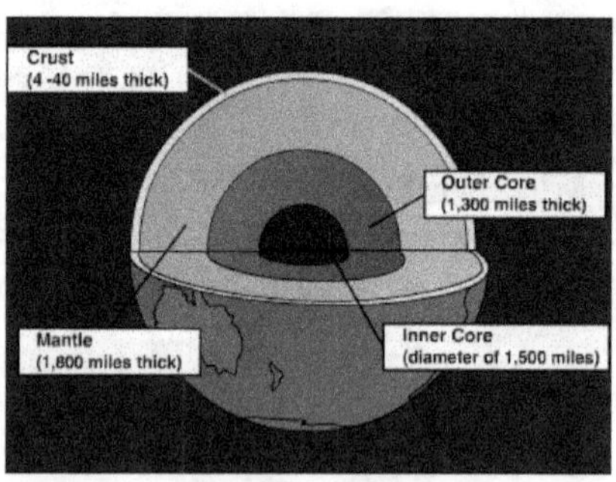

Image: The Hot Zone

The outermost layer of the Earth is the crust. In the oceans, it is about 4 miles thick. On the thickest continents, it extends some 40 miles into the Earth. Below the crust is the mantle, a layer of weaker, hotter rock some 1,800 miles thick. Beneath the mantle lies the outer core, a sea of liquid iron about 1,300 miles thick. Finally, at the center of the Earth is the solid iron core, a sphere some 1,500 miles across.

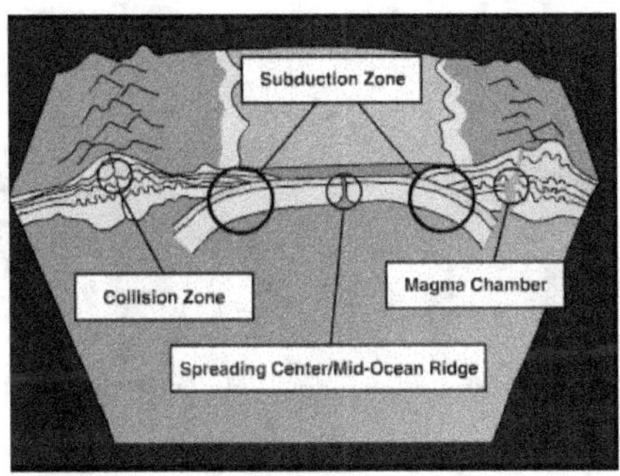

This illustration shows the surface of the Earth and the upper part of its interior, and the forces that produce the violent activity we see on the surface. (It does not represent a picture of any real geographical area, but is meant to illustrate a number of different geological processes.)

Shown at left is a collision zone where continental plates collide, as they do, for example, in the Asian subcontinent. At either edge of the ocean (blue) are subduction zones, where the edges of the heavier oceanic plate are slowly sinking below continental plates. Subduction at the oceanic plate's edges very gradually pulls the plate apart in the middle, forming what is called a "spreading center" or "mid-ocean ridge," shown at middle.

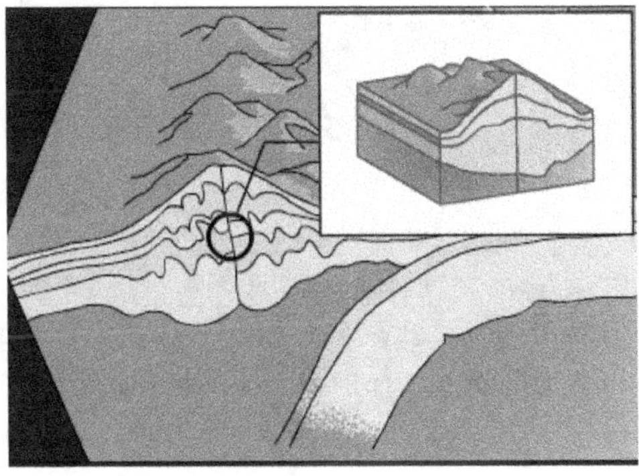

Where continental plates collide, they crunch together and form mountains and cause earth quake.

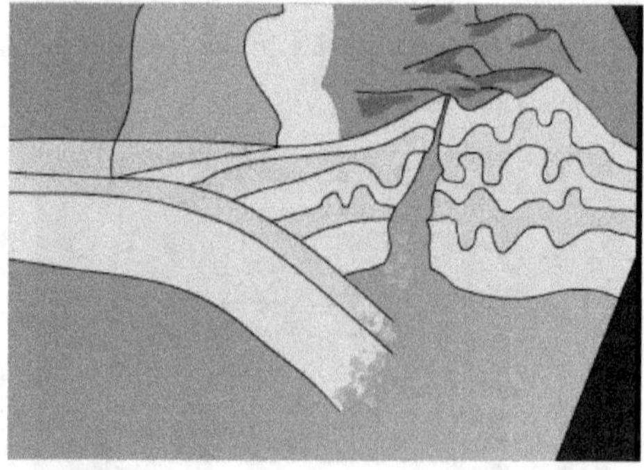

When the edge of a cold, rocky oceanic plate sinks in to the mantle under a continental plate, it is heated and deformed and becomes part of the mantle. Large earthquakes often occur in these areas. Volcanoes also form within the continental plate when melting occurs in the mantle above the sub -ducting oceanic plate, feeding "magma chambers" within the crust. The magma can find its way to the surface as volcanoes.

As its edges sub duct in different directions, the oceanic plate is pulled apart, in the middle, forming a "spreading center" or "mid ocean ridge". Magma percolates up in the middle and cools to form new crust (which gradually sinks to the sides forming a ridge). The birth of new crust is accompanied by earth quakes. (Not shown: Volcanic Island like Surtsey and Heimaey, off the coast of Iceland and featured in HELL'S CRUST, may also form)

References

1. T. H. Jordan (1979). "Structural Geology of the Earth's Interior". Proceedings of the National Academy of Sciences 76 (9):4192–4200. Bibcode:1979PNAS..76.4192J. doi: 10.1073/pnas.76.9.4192. PMC 411539. PMID 16592703.

2. Monnereau, Marc; Calvet, Marie; Margerin, Ludovic; Souriau, Annie (May 21, 2010). "Lopsided Growth of Earth's Inner Core".Science 328 (5981): 1014–1017. Bibcode: 2010Sci...328.1014M. doi:10.1126/science.1186212. PMID 20395477.

3. Hazlett, James S. Monroe; Reed Wicander; Richard (2006). Physical geology : exploring the earth; [the wrath of Hurricane Katrina ; Could you survive a Tsunami?; catastrophic earthquakes; global warming](6.ed.). Belmont: Thomson. P. 346. ISBN 9780495011484

4. Uwe Walzer, Roland Hendel, John Baumgardner Mantle Viscosity and the Thickness of the Convective Downwellings.

5. Breaking News | Oldest rock shows Earth was a hospitable young planet. Spaceflight Now (2001-01-14). Retrieved on 2012-01-27

6. http://www.bbc.co.uk/schools/gcsebitesize/geography/natural_hazards/tectonic_plates_rev1.shtm

7. http://www.bbc.co.uk/schools/gcsebitesize/geography/natural_hazards/tectonic_plates_rev2.shtml

8. http://www.bbc.co.uk/schools/gcsebitesize/geography/natural_hazards/tectonic_plates_rev3.shtml

9. http://www.bbc.co.uk/schools/gcsebitesize/geography/natural_hazards/tectonic_plates_rev3.shtml

10. http://solarviews.com/eng/earthint.htm

11. http://solarviews.com/raw/earth/earthfg2.gif

12. Beatty, J. K. and A. Chaikin, eds. The New Solar System. Massachusetts: Sky Publishing, 3rd Edition, 1990.

13. Press, Frank and Raymond Siever. Earth. New York: W. H. Freeman and Company, 1986.

14. Seeds, Michael A. Horizons. Belmont, California: Wadsworth, 1995.

15. http://autoreview2016.xyz/tag/inner-core neol01 img0027

16. https://en.wikipedia.org/wiki/Inner_core

17. http://ase.tufts.edu/cosmos/print_images.asp?id=4

18. http://www.pbs.org/wnet/savageearth/animations/hellscrust/main.html

Resources

1. M. Dziewonski, D. L. Anderson (1981). "Preliminary reference Earth model"(PDF). Physics of the Earth and Planetary Interiors 25 (4): 297–356. doi:10.1016/0031-9201(81)90046-7. ISSN 0031-9201.

2. Stixrude, Lars; Cohen, R.E. (January 15, 1995). "Constraints on the crystalline structure of the inner core: Mechanical instability of BCC iron at high pressure". Geophysical Research Letters 22 (2):125–128. Bibcode:1995GeoRL..22..125S. doi:10.1029/94GL02742.

3. Benuzzi-Mounaix, A.; Koenig, M.; Ravasio, A.; Vinci, T. (2006). "Laser-driven shock waves for the study of extreme matter states". Plasma Physics and Controlled Fusion 48 (12B). Bibcode:2006PPCF...48B.347B. doi:10.1088/0741-3335/48/12B/S32.

4. Remington, Bruce A.; Drake, R. Paul; Ryutov, Dmitri D.(2006). "Experimental astrophysics with high power lasers and Z pinches". Reviews of Modern Physics 78. Bibcode:2006RvMP...78..755R. doi:10.1103/RevModPhys.78.755.

5. Benuzzi-Mounaix, A.; Koenig, M.; Husar, G.; Faral, B. (June 2002). "Absolute equation of state measurements of iron using laser driven shocks". Physics of Bibcode: 2002PhPl.... 9.2466B. doi:10.1063/1.1478557.

6. Cohen, Ronald; Stixrude, Lars. "Crystal at the Center of the Earth". Retrieved 2007-02-05.

7. Stixrude, L.; Cohen, R. E. (1995). "High-Pressure Elasticity of Iron and Anisotropy of Earth's Inner Core". Science 267 (5206):1972–5. Bibcode:1995Sci...267.1972S. doi: 10. 1126/science.267.5206.1972. PMID 17770110.

8. BBC News, "What is at the centre of the Earth?. Bbc.co.uk (2011-08-31). Retrieved on 2012-01-27.

9. Ozawa, H.; al., et (2011). "Phase Transition of FeO and Stratification in Earth's Outer Core". Science 334 (6057):792–794. Bibcode:2011Sci...334..792O. doi:10.1126/science. 1208265.

10. Wootton, Anne (2006). "Earth's Inner Fort Knox". Discover 27 (9): 18.

11. Herndon, J. M. (1980). "The chemical composition of the interior shells of the Earth". Proc. R. Soc. Lond A372 (1748): 149–154. JSTOR 2398362.

12. J. M. (2005). "Scientific basis of knowledge on Earth's composition" (PDF). Current Science 88 (7): 1034–1037.

13. First Measurement Of Magnetic Field Inside Earth's Core. Science20.com. Retrieved on 2012-01-27.

14. Buffett, Bruce A. (2010). "Tidal dissipation and the strength of the Earth's internal magnetic field". Nature 468 (7326):952–4. Bibcode:2010Natur.468..952B. doi:10.1038/nature09643. PMID 21164483.

15. Chang, Kenneth (2005-08-25). "Earth's Core Spins Faster Than the Rest of the Planet". The New York Times. Retrieved2010-05-24.

16. Kerr, R. A. (2005). "Earth's Inner Core Is Running a Tad Faster Than the Rest of the Planet". Science 309 (5739):1313a.doi:10.1126/science.309.5739.1313a. PMID 16123276.

17. Chang, Kenneth (26 August 2005) "Scientists Say Earth's Center Rotates Faster Than Surface" The New York Times Sec. A, Col. 1, p. 13

18. N. Kollerstrom (1992). "The hollow world of Edmond Halley". Journal for History of Astronomy 23: 185–192. Archive.

4. The Science of Earthquakes

4.1. Seismology and Seismic Waves [1]

Seismology is the study of earthquakes and seismic waves that move through and around the earth.

Seismic waves are the waves of energy caused by the sudden breaking of rock within the earth or an explosion. These are the energy that travels through the earth and is recorded on seismographs.

There are several different kinds of seismic waves, and they all move in different ways. The two main types of waves are **body waves** and **surface waves**. Body waves can travel through the earth's inner layers, but surface waves can only move along the surface of the planet like ripples on water. Earthquakes radiate seismic energy as both body and surface waves.

Traveling through the interior of the earth, **body waves** arrive before the surface waves emitted by an earthquake. These waves are of a higher frequency than surface waves.

4.1.1. Body Waves

- *P Waves [2]*

The first kind of body wave is the **P wave** or **primary wave**. This is the fastest kind of seismic wave, and, consequently, the first to 'arrive' at a seismic station. The P wave can move through solid rock and fluids, like water or the liquid layers of the earth. It pushes and pulls the rock it moves through just like sound waves push and pull the air. Have you ever heard a big clap of thunder and heard the windows rattle at the same time? The windows rattle because the sound waves were pushing and pulling on the window glass much like P waves push and pull on rock. Sometimes animals can hear the P waves of an earthquake. Dogs, for instance, commonly begin barking hysterically just before an earthquake 'hits' (or more specifically, before the surface waves arrive). Usually people can only feel the bump and rattle of these waves. P waves are also known as **compression waves**, because of the pushing and pulling they do. Subjected to a P wave, particles move in the same direction that the wave is moving in, which is the direction that the energy is traveling in, and is sometimes called the 'direction of wave propagation'.

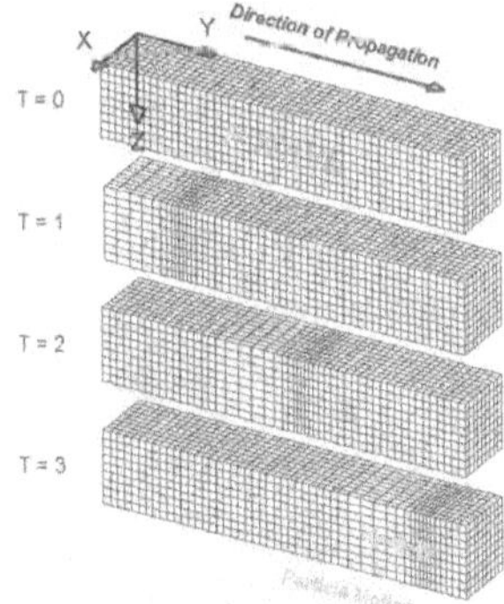

Figure 1: A P Wave Travels through a Medium by Means of Compression and Dilation. Particles are represented by cubes in this Model

Image: ©2000-2006 Lawrence Braile, used with permission.

- ### *S Waves* [3]

The second type of body wave is the S wave or secondary wave, which is the second wave you feel in an earthquake. An S wave is slower than a P wave and can only move through solid rock, not through any liquid medium. It is this property of S waves that led seismologists to conclude that the Earth's outer core is a liquid. S waves move rock particles up and down, or side-to-side--perpendicular to the direction that the wave is traveling in (the direction of wave propagation).

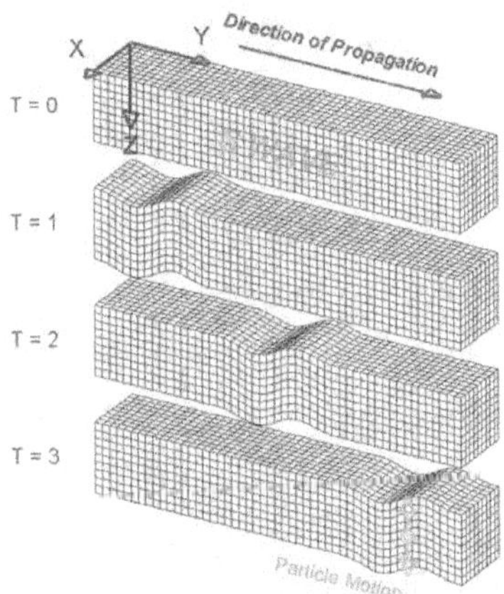

Figure 2: An S Wave Travels through a Medium. Particles are represented by Cubes in this Model. Image: ©2000-2006 Lawrence Braile, Used with Permission

4.1.2. Surface Waves

Travelling only through the crust, **surface waves** are of a lower frequency than body waves, and are easily distinguished on a seismogram as a result. Though they arrive after body waves, it is surface waves that are almost enitrely responsible for the damage and destruction associated with earthquakes. This damage and the strength of the surface waves are reduced in deeper earthquakes.

- ### Love Waves [4]

The first kind of surface wave is called a **Love wave**, named after A.E.H. Love, a British mathematician who worked out the mathematical model for this kind of wave in 1911. It's the fastest surface wave and moves the ground from side-to-side. Confined to the surface of the crust, Love waves produce entirely horizontal motion.

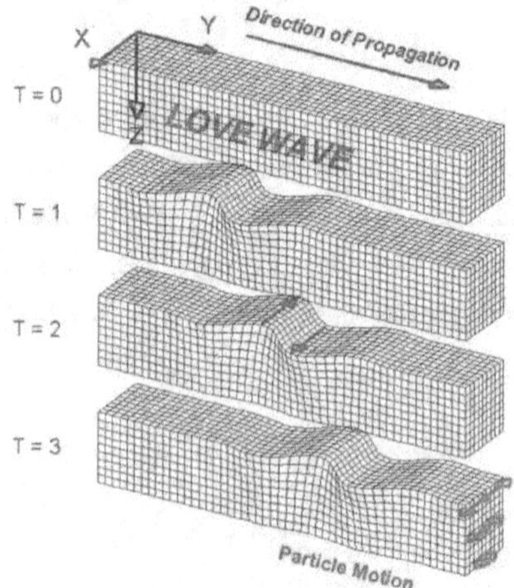

Figure 3: A Love Wave Travels through a Medium. Particles are represented by Cubes in this Model. Image: ©2000-2006 Lawrence Braile, Used with Permission

- ### Rayleigh Waves [5]

The other kind of surface wave is the **Rayleigh wave**, named for John William Strutt, Lord Rayleigh, who mathematically predicted the existence of this kind of wave in 1885. A Rayleigh wave rolls along the ground just like a wave rolls across a lake or an ocean. Because it rolls, it moves the ground up and down and side-to-side in the same direction that the wave is moving. Most of the shaking felt from an earthquake is due to the Rayleigh wave, which can be much larger than the other waves.

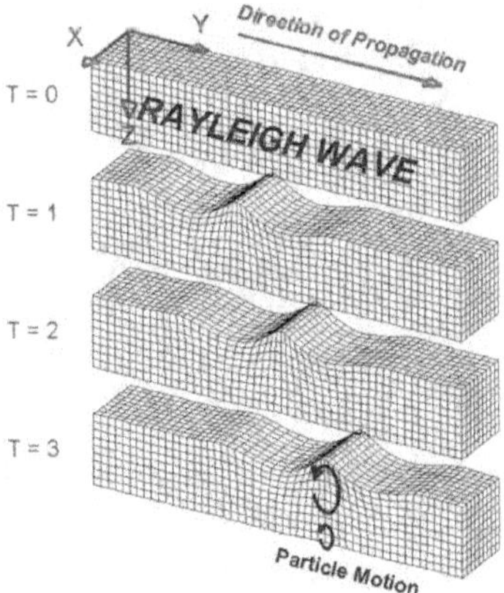

Figure 4: A Rayleigh wave travels through a medium. Particles are represented by cubes in this model. Image: ©2000-2006 Lawrence Braile, used with permission.

4.2.   Earthquake Occurrence

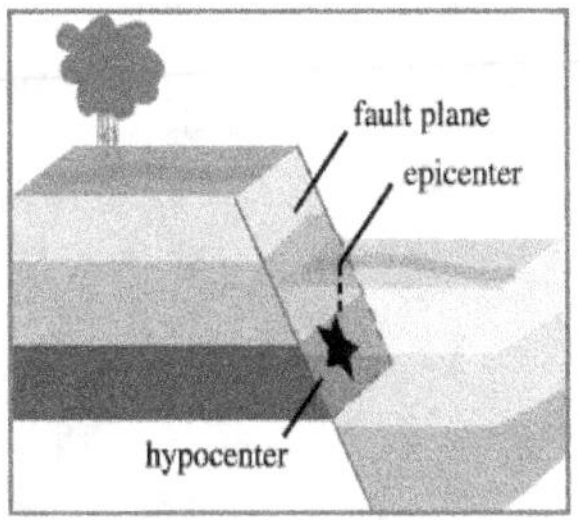

Figure 5

As shown in (Figure 5), the earthquake happens when two blocks of the earth suddenly slip past one another. The surface where they slip is called the *fault* or *fault plane*. The location below the earth's surface where the earthquake starts is called the *hypocenter*, and the location directly above it on the surface of the earth is called the *epicenter*.

Sometimes an earthquake has *foreshocks*. These are smaller earthquakes that happen in the same place as the larger earthquake that follows. Scientists can't tell that an earthquake is a foreshock until the larger earthquake happens. The largest, main earthquake is called the *main shock*. Main shocks always have *aftershocks* that follow. These are smaller earthquakes that occur afterwards in the same place as the main shock. Depending on the size of the main shock, aftershocks can continue for weeks, months, and even years after the main shock!

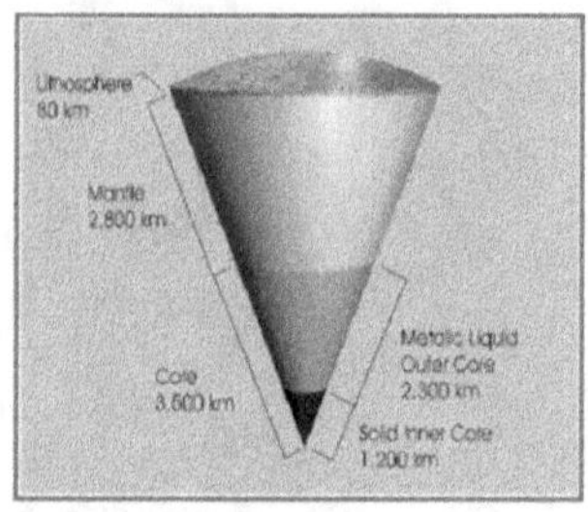

Figure 6

The earth has four major layers: the *inner core, outer core, mantle* and *crust* (Figure 6).The crust and the top of the mantle make up a thin skin on the surface of our planet. But this skin is not all in one piece – it is made up of many pieces like a puzzle covering the surface of the earth (Figure 7). Not only that, but these puzzle pieces keep slowly moving around, sliding past one another and bumping into each other. We call these puzzle pieces *tectonic plates*, and the edges of the plates are called the *plate* boundaries. The plate boundaries are made up of many faults, and most of the earthquakes around the world occur on these faults. Since the edges of the plates are rough, they get stuck while the rest of the plate keeps moving. Finally, when the plate has moved far enough, the edges unstuck on one of the faults and there is an earthquake.

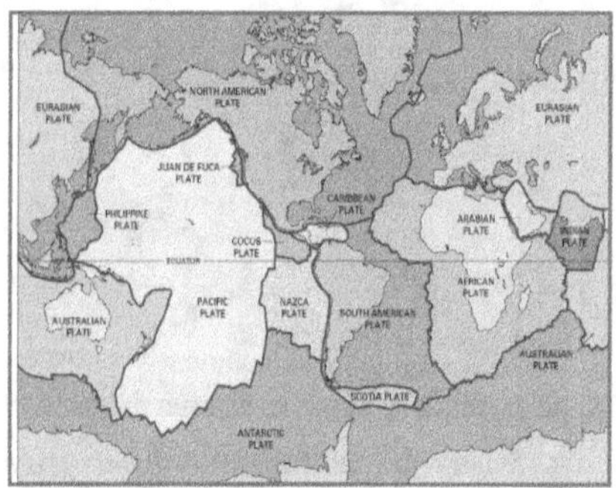

Figure 7

While the edges of faults are stuck together, and the rest of the block is moving, the energy that would normally cause the blocks to slide past one another is being stored up. When the force of the moving blocks finally overcomes the *friction* of the jagged edges of the fault and it unstuck, all that stored up energy is released. The energy radiates outward from the fault in all directions in the form of *seismic waves* like ripples on a pond. The seismic waves shake the earth as they move through it, and when the waves reach the earth's surface, they shake the ground and anything on it, like our houses and us!

4.3.   Recording Earthquake

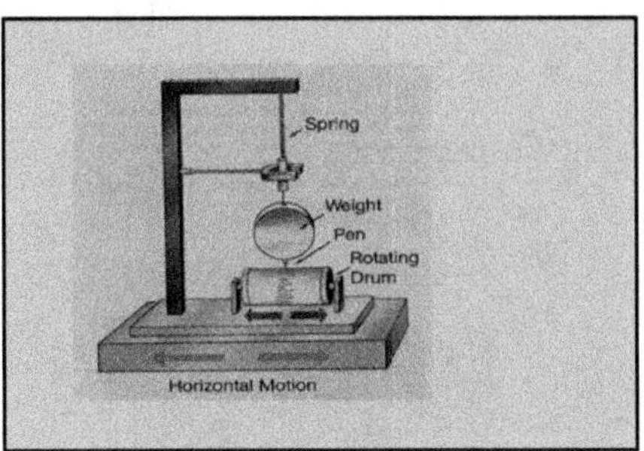

Figure 8: Recording Earthquake

Earthquakes are recorded by instruments called *seismographs* (Figure 8) and the recording made is called a *seismogram.* The seismograph has a base that sets firmly in the ground, and a heavy weight that hangs free. When an earthquake causes the ground to shake, the base of the seismograph shakes too, but the hanging weight does not. Instead, the spring or string from that the weight is hanging absorbs all the movement. The difference in position between the shaking part of the seismograph and the motionless part is what is recorded.

4.4.   Measurement of the Size of Earthquakes

The size of an earthquake depends on the size of the fault and the amount of slip on the fault, but that's not something scientists can simply measure with a measuring tape since faults are many kilometers deep beneath the earth's surface. So for measuring an earthquake, they use the *seismogram* recordings made on the *seismographs* at the surface of the earth to determine how large the earthquake was (Figure 9). A short wiggly line that doesn't wiggle very much means a small earthquake, and a long wiggly line that wiggles a lot means a large earthquake. The length of the wiggle depends on the size of the fault, and the size of the wiggle depends on the amount of slip.

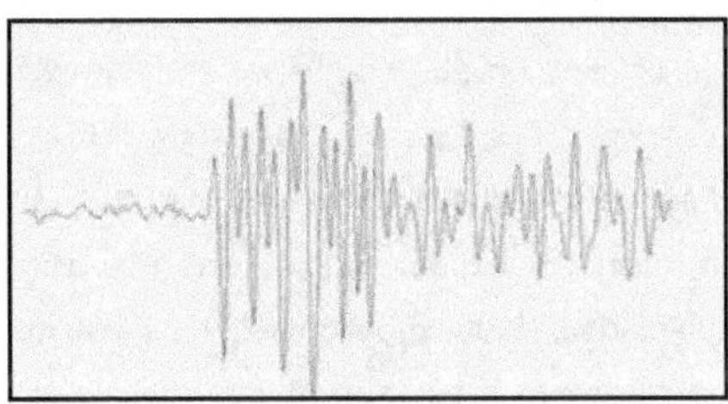

Figure 9: A Typical Seismic Wave

The size of the earthquake is called its *magnitude*. There is one magnitude for each earthquake. Scientists also talk about the *intensity* of shaking from an earthquake, and this varies depending on where one is during the earthquake.

4.5.   Locating the Earthquake

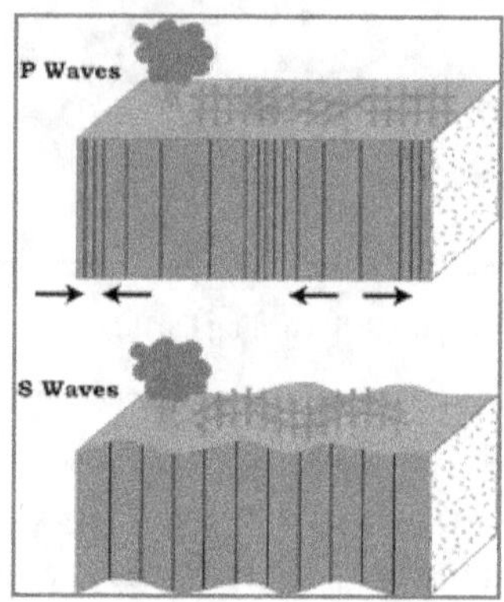

Seismograms come in handy for locating earthquakes too and being able to see the *P wave* and the *S wave* which is important. P & S waves each shake the ground in different ways as they travel through it. P waves are also faster than S waves, and this fact is what allows us to tell where an earthquake was. To understand how this works, let's compare P and S waves to lightning and thunder. Light travels faster than sound, so during a thunderstorm one will first see the lightning and then hear the thunder. If you are close to the lightning, the thunder will boom right after the lightning, but if you are far away from the lightning, you can count several seconds before you hear the thunder. The farther you are from the storm, the longer it will take between the lightning and the thunder.

P waves are like the lightning, and S waves are like the thunder. The P waves travel faster and shake the ground where you are first. Then the S waves follow and shake the ground also. If you are close to the earthquake, the P and S wave will come one right after the other, but if you are far away, there will be more time interval between the two. By looking at the extent of time interval between the arrival of P and S wave on a seismogram recorded on a seismograph, scientists can tell how far away the earthquake was from that location. However, they can't tell in what direction from the seismograph the earthquake was. They can only say how far away it was from the seismograph station. If they draw a circle on a map around the station where the *radius* of the circle is the determined distance to the earthquake, they know the earthquake lies somewhere on the circle. But where on the circle, is not yet known.

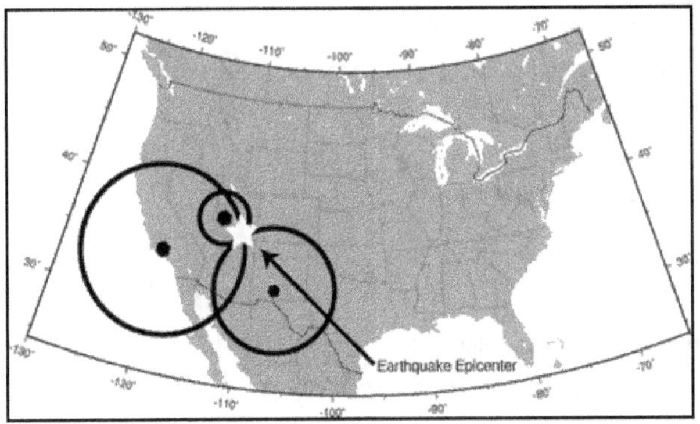

Figure 10: Epicenter

Scientists then use a method called *triangulation* to determine exactly where the earthquake was (Figure 10). It is called triangulation because a triangle has three sides, and it takes three seismographs to locate an earthquake. If you draw a circle on a map around three different seismographs where the *radius* of each is the distance from that station to the earthquake, the point of intersection of those three circles is the *epicenter*!

Thus the actual location of the earthquake's epicenter will be on the perimeter of a circle drawn around the recording **station**. The radius of this circle is the epicentral distance. One S - P measurement will produce one epicentral distance: the direction from which the waves came is unknown.

4.6.   Measuring and Locating Earthquakes

Seismology

Earthquakes can be recorded by seismometers up to great distances, because seismic waves travel through the whole Earth's interior. The absolute magnitude of a quake is conventionally reported by numbers on the moment magnitude scale (formerly Richter scale, magnitude 7 causing serious damage over large areas), whereas the felt magnitude is reported using the modified Mercalli intensity scale (intensity II–XII).

Every tremor produces different types of seismic waves, which travel through rock with different velocities:

- Longitudinal P-waves (shock- or pressure waves)
- Transverse S-waves (both body waves)
- Surface waves -(Rayleigh and Love waves)

Propagation velocity of the seismic waves ranges from approx. 3 km/s up to 13 km/s, depending on the density and elasticity of the medium. In the Earth's interior the shock- or P waves travel much faster than the S waves **(approx. relation 1.7:1)**. The differences in travel time from the epicenter to the observatory are a measure of the distance and can be used to image both sources of quakes and structures within the Earth. Also the depth of the hypocenter can be computed roughly.

In solid rock P-waves travel at about 6 to 7 km per second; the velocity increases within the deep mantle to ~13 km/s. The velocity of S-waves ranges from 2–3 km/s in light sediments and 4–5 km/s in the Earth's crust up to 7 km/s in the deep mantle. As a consequence, the first waves of a distant earthquake arrive at an observatory via the Earth's mantle.

On average, the kilometer distance to the earthquake is the number of seconds between the P and S wave **times 8**.[6]Slight deviations are caused by inhomogeneities of subsurface structure. By such analyses of seismograms the Earth's core was located in 1913 by Beno Gutenberg.

Earthquakes are not only categorized by their magnitude but also by the place where they occur. The world is divided into 754 Flinn–Engdahl regions (F-E regions), which are based on political and geographical boundaries as well as seismic activity. More active zones are divided into smaller F-E regions whereas less active zones belong to larger F-E regions.

Standard reporting of earthquakes includes its magnitude, date and time of occurrence, geographic coordinates of its epicenter, depth of the epicenter, geographical region, distances to population centers, location uncertainty, a number of parameters that are included in USGS earthquake reports (number of stations reporting, number of observations, etc.), and a unique event ID.[7]

To figure out just where that earthquake happened, it is necessary to look at the seismogram and to know about at least two other seismographs recorded for the same earthquake.

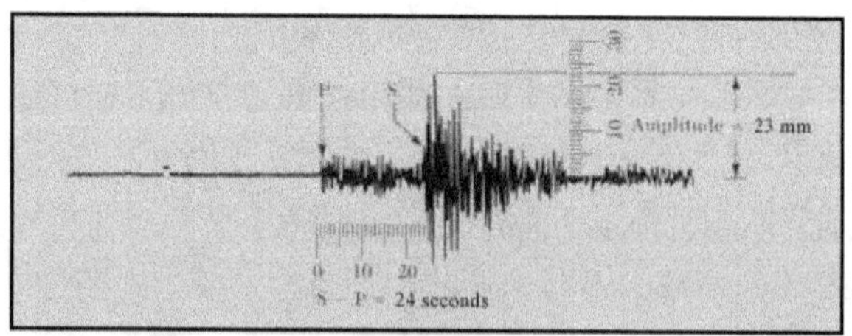

Figure 11: A Typical Example of a Seismogram: (from bolt, 1978).

In the Figure 5, one minute intervals are marked by the small lines printed just above the squiggles made by the seismic waves (the time may be marked differently on some seismographs). The distance between the beginning of the first P wave and the first S wave tells how many seconds the waves are apart. This number is used to know how far your seismograph is from the epicenter of the earthquake.

4.6.1. Finding the Distance to the Epicenter and the Earthquake's Magnitude

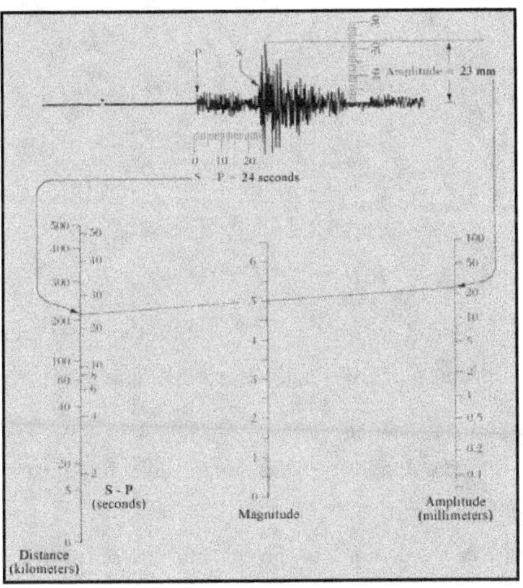

Figure 12: The Amplitude is Used to derive the Magnitude of the Earthquake, and the Distance from the Earthquake to the Station. (from bolt, 1978)

Steps

1. Measure the distance between the first P wave and the first S wave. In this case, the first P and S waves are 24 seconds apart.

2. Find the point for 24 seconds on the left side of the chart and mark that point. According to the chart, this earthquake's epicenter was 215 kilometers away.

3. Measure the amplitude of the strongest wave. The amplitude is the height (on paper) of the strongest wave. On this seismogram, the amplitude is 23 millimeters. Find 23 millimeters on the right side of the chart and mark that point.

4. Place a ruler (or straight edge) on the chart between the points you marked for the distance to the epicenter and the amplitude. The point where your ruler crosses the middle line on the chart marks the magnitude (strength) of the earthquake. This earthquake had a magnitude of 5.0.

4.6.2. *Finding the Epicenter*

You have just figured out how far your seismograph is from the epicenter and how strong the earthquake was, but you still don't know exactly where the earthquake occurred. This is where the compass, the map, and the other seismograph records come in.

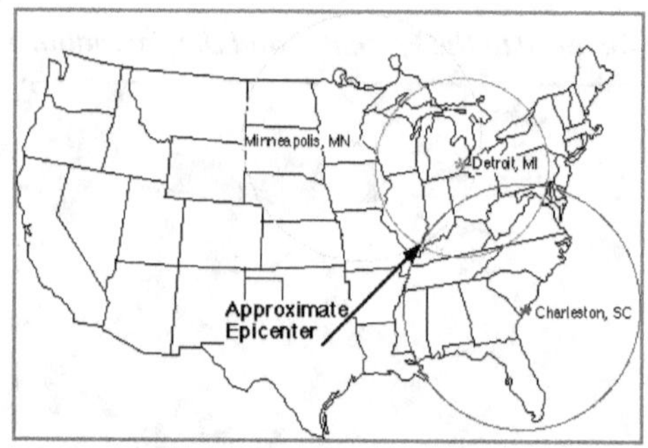

Figure 13: The Point where the three Circles Intersect is the Epicenter of the Earthquake. This Technique is called 'Triangulation.'

- Check the scale on your map. It should look something like a piece of a ruler. All maps are different. On your map, one centimeter could be equal to 100 kilometers or something like that.

- Figure out how long the distance to the epicenter (in centimeters) is on your map. For example, say your map has a scale where one centimeter is equal to 100 kilometers. If the epicenter of the earthquake is 215 kilometers away, that equals 2.15 centimeters on the map.

- Using your compass, draw a circle with a radius equal to the number you came up with in Step #2 (the **radius** is the distance from the center of a circle to its edge). The center of the circle will be the location of your seismograph. The epicenter of the earthquake is somewhere on the edge of that circle (i.e. the actual location of the earthquake's epicenter will be on the perimeter of a circle drawn around the recording station). The radius of this circle is the epicentral distance. One S - P measurement will produce one epicentral distance: the direction from which the waves came is unknown.

- Do the same thing for the distance to the epicenter that the other seismograms recorded (with the location of those seismographs at the center of their circles). All of the circles should overlap. The point where all of the circles overlap is the approximate epicenter of the earthquake.

4.7.   Prediction of Earthquakes

Till today it has not been possible to predict earthquakes and it is unlikely that it can ever be done. Scientists have tried many different ways of predicting earthquakes, but none have been successful. On any particular fault, scientists know there will be another earthquake sometime in the future, but they have no way of telling when it will happen.

Weather does not affect earthquake occurrence; animals or people also do not exhibit any sign. These are two questions that do not yet have definite answers. If weather does affect earthquake occurrence, or if some animals or people can tell when an earthquake is coming, it is not yet understood how it works.

4.8.   Naturally Occurring Earthquakes

4.8.1. Fault Types

Tectonic earthquakes occur anywhere in the earth where there is sufficient stored elastic strain energy to drive fracture propagation along a fault plane. The sides of a fault move past each other smoothly and a seismically only if there are no irregularities or asperities along the fault surface that increase the frictional resistance. Most fault surfaces do have such asperities and this leads to a form of stick-slip behavior. Once the fault has locked, continued relative motion between the plates leads to increasing stress and therefore, stored strain energy in the volume around the fault surface. This continues until the stress has risen sufficiently to break through the asperity, suddenly allowing sliding over the locked portion of the fault, releasing the stored energy.[8] This energy is released as a combination of radiated elastic strain seismic waves, frictional heating of the fault surface, and cracking of the rock, thus causing an earthquake. This process of gradual build-up of strain and stress punctuated by occasional sudden earthquake failure is referred to as the elastic-rebound theory. It is estimated that only 10 percent or less of an earthquake's total energy is radiated as seismic energy. Most of the earthquake's energy is used to power the earthquake fracture growth or is converted into heat generated by friction. Therefore, earthquakes lower the Earth's available elastic potential energy and raise its temperature, though these changes are negligible compared to the conductive and convective flow of heat out from the Earth's deep interior.[9]

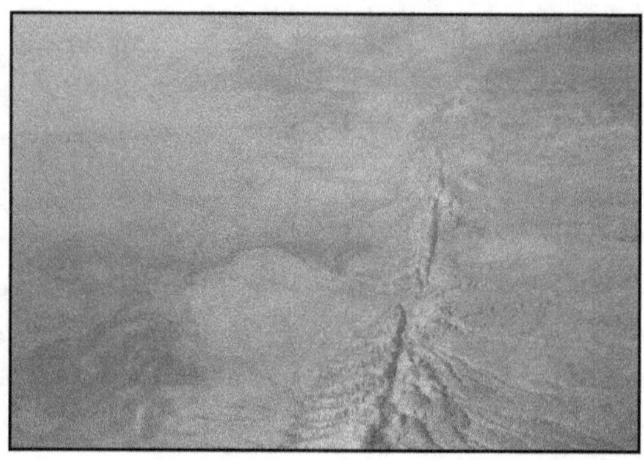

Figure 14: Aerial Photo of the San Andreas Fault in the Carrizo Plain, northwest of Los Angeles

There are three main types of fault, all of which may cause an interplate earthquake: normal, reverse (thrust) and strike-slip. Normal and reverse faulting are examples of dip-slip, where the displacement along the fault is in the direction of dip and movement on them involves a vertical component. Normal faults occur mainly in areas where the crust is being extended such as a divergent boundary. Reverse faults occur in areas where the crust is being shortened such as at a convergent boundary. Strike-slip faults are steep structures where the two sides of the fault slip horizontally past each other; transform boundaries are a particular type of strike-slip fault. Many earthquakes are caused by movement on faults that have components of both dip-slip and strike-slip; this is known as oblique slip.

Reverse faults, particularly those along convergent plate boundaries are associated with the most powerful earthquakes, mega thrust earthquakes, including almost all of those of magnitude 8 or more. Strike-slip faults, particularly continental transforms, can produce major earthquakes up to about magnitude 8. Earthquakes associated with normal faults are generally less than magnitude 7. For every unit increase in magnitude, there is a roughly thirtyfold increase in the energy released. For instance, an earthquake of magnitude 6.0 releases approximately 30 times more energy than a 5.0 magnitude earthquake and a 7.0 magnitude earthquake releases 900 times (30 × 30) more energy than a 5.0 magnitude of earthquake. An 8.6 magnitude earthquake releases the same amount of energy as 10,000 atomic bombs like those used in World War II.[10]

This is so because the energy released in an earthquake, and thus its magnitude, is proportional to the area of the fault that ruptures[11] and the stress drop. Therefore, the longer the length and the wider the width of the faulted area, the larger the resulting magnitude. The topmost, brittle part of the Earth's crust, and the cool slabs of the tectonic plates that are descending down into the hot mantle, are the only parts of our planet which can store elastic energy and release it in fault ruptures. Rocks hotter than about 300 degrees Celsius flow in response to stress; they do not rupture in earthquakes.[12][13]The maximum observed lengths of ruptures and mapped faults (which may break in a single rupture) are approximately 1000 km. Examples are the earthquakes in Chile, 1960; Alaska, 1957; Sumatra, 2004, all in sub duction zones. The longest earthquake ruptures on strike-slip faults, like the San Andreas Fault (1857, 1906), the North Anatolian Fault in Turkey (1939) and the Denali Fault in Alaska (2002), are about half to one third as long as the lengths along subducting plate margins, and those along normal faults are even shorter.

The most important parameter controlling the maximum earthquake magnitude on a fault is however not the maximum available length, but the available width because the latter varies by a factor of 20. Along converging plate margins, the dip angle of the rupture plane is very shallow, typically about 10 degrees.[14] Thus the width of the plane within the top brittle crust of the Earth can become 50 to 100 km making the most powerful earthquakes possible.

Strike-slip faults tend to be oriented near vertically, resulting in an approximate width of 10 km within the brittle crust,[15] thus earthquakes with magnitudes much larger than 8 are not possible. Maximum magnitudes along many normal faults are even more limited because many of them are located along spreading centers, as in Iceland, where the thickness of the brittle layer is only about 6 km.[16][17]

In addition, there exists a hierarchy of stress level in the three fault types. Thrust faults are generated by the highest, strike slip by intermediate, and normal faults by the lowest stress levels.[18] This can easily be understood by considering the direction of the greatest principal stress, the direction of the force that 'pushes' the rock mass during the faulting. In the case of normal faults, the rock mass is pushed down in a vertical direction, thus the pushing force (**greatest** principal stress) equals the weight of the rock mass itself. In the case of thrusting, the rock mass 'escapes' in the direction of the least principal stress, namely upward, lifting the rock mass up, thus the overburden equals the **least** principal stress. Strike-slip faulting is intermediate between the other two types described above. This difference in stress regime in the three faulting environments can contribute to differences in stress drop during faulting, which contributes to differences in the radiated energy, regardless of fault dimensions.

4.8.2. Earthquakes away from Plate Boundaries

Intraplate Earthquake

Where plate boundaries occur within the continental lithosphere, deformation is spread out over a much larger area than the plate boundary itself. All tectonic plates have internal stress fields caused by their interactions with neighboring plates and sedimentary loading or unloading (e.g. deglaciation).[19] These stresses may be sufficient to cause failure along existing fault planes, giving rise to intraplate earthquakes.[20]

4.8.3. Shallow-focus and Deep-focus Earthquakes

Depth of Focus (Tectonics)

The majority of tectonic earthquakes originate at the ring of fire in depths not exceeding tens of kilometers. Earthquakes occurring at a depth of less than 70 km are classified as 'shallow-focus' earthquakes, while those with a focal-depth between 70 and 300 km are commonly termed 'mid-focus' or 'intermediate-depth' earthquakes.

In subduction zones, where older and colder oceanic crust descends beneath another tectonic plate, deep-focus earthquakes may occur at much greater depths(ranging from 300 upto 700 kilometers).[21] Deep-focus earthquakes occur at a depth where the subducted lithosphere should no longer be brittle, due to the high temperature and pressure. A possible mechanism for the generation of deep-focus earthquakes is faulting caused by olivine undergoing a phase transition into a spinel structure. [22]

4.8.4. Earthquakes and Volcanic Activity

Earthquakes often occur in volcanic regions and are caused there, both by tectonic faults and the movement of magma in volcanoes. Such earthquakes can serve as an early warning of volcanic eruptions,.[23] Earthquake swarms can serve as markers for the location of the flowing magma throughout the volcanoes. These swarms can be recorded by seismometers and tiltmeters (a device that measures ground slope) and used as sensors to predict imminent or upcoming eruptions.[24]

4.8.5. Rupture Dynamics

A tectonic earthquake begins by an initial rupture at a point on the fault surface, a process known as nucleation. The scale of the nucleation zone is uncertain, with some evidence, such as the rupture dimensions of the smallest earthquakes, suggesting that it is smaller than 100 m while other evidence, such as a slow component revealed by low-frequency spectra of some earthquakes, suggest that it is larger. The possibility that the nucleation involves some sort of preparation process is supported by the observation that about 40% of earthquakes are preceded by foreshocks. Once the rupture has initiated, it begins to propagate along the fault surface. The mechanics of this process are poorly understood, partly because it is difficult to recreate the high sliding velocities in a laboratory. Also the effects of strong ground motion make it very difficult to record information close to a nucleation zone.[25]

Rupture propagation is generally modeled using a fracture mechanics approach, likening the rupture to a propagating mixed mode shear crack. The rupture velocity is a function of the fracture energy in the volume around the crack tip, increasing with decreasing fracture energy. The velocity of rupture propagation is orders of magnitude faster than the displacement velocity across the fault. Earthquake ruptures typically propagate at velocities that are in the range 70–90% of the S-wave velocity, and this is independent of earthquake size. A small subset of earthquake ruptures appear to have propagated at speeds greater than the S-wave velocity. These supershear earthquakes have all been observed during large strike-slip events. Some earthquake ruptures travel at unusually low velocities and are referred to as slow earthquakes. A particularly dangerous form of slow earthquake is the tsunami earthquake, observed where the relatively low felt intensities, caused by the slow propagation speed of some great earthquakes, fail to alert the population of the neighbouring coast,.[25]

4.8.6. Tidal Forces

Research work has shown a robust correlation between small tidally induced forces and non-volcanic tremor activity. [26] [27] [28] [29]

4.8.7. Earthquake Clusters

Most earthquakes form part of a sequence, related to each other in terms of location and time.[30] Most earthquake clusters consist of small tremors that cause little to no damage, but there is a theory that earthquakes can recur in a regular pattern.[31]

4.8.8. Aftershock

An aftershock is an earthquake that occurs after a previous earthquake, the mainshock. An aftershock is in the same region of the main shock but always of a smaller magnitude. If an aftershock is larger than the main shock, the aftershock is redesignated as the main shock and the original main shock is redesignated as a foreshock. Aftershocks are formed as the crust around the displaced fault plane adjusts to the effects of the main shock.[30]

4.8.9. *Earthquake Swarm*

Earthquake swarms are sequences of earthquakes striking in a specific area within a short period of time. They are different from earthquakes followed by a series of aftershocks by the fact that no single earthquake in the sequence is obviously the main shock; therefore none have notable higher magnitudes than the other.

4.8.10. *Earthquake Storm*

Sometimes a series of earthquakes occur in a sort of earthquake storm, where the earthquakes strike a fault in clusters, each triggered by the shaking or stress redistribution of the previous earthquakes. Similar to aftershocks but on adjacent segments of fault, these storms occur over the course of years, and with some of the later earthquakes as damaging as the early ones

4.8.11. *Size and Frequency of Occurrence*

It is estimated that around 500,000 earthquakes occur each year, detectable with current instrumentation. About 100,000 of these can be felt.[32][33] Minor earthquakes occur nearly constantly around the world.

4.8.12. *Induced Seismicity*

While most earthquakes are caused by movement of the Earth's tectonic plates, human activity can also produce earthquakes. Four main activities contribute to this phenomenon:

1. *Storing large amounts of water behind a dam (and possibly building an extremely heavy building),*
2. *Drilling and injecting liquid into wells*
3. *Coal mining*
4. *Oil drilling.[34]*

References

1. http://www.geo.mtu.edu/UPSeis/waves.html
2. http://www.geo.mtu.edu/upseis/images/p-wave_animation.gif
3. http://www.geo.mtu.edu/upseis/images/s-wave_animation.gif
4. http://www.geo.mtu.edu/upseis/images/love_animation.gif
5. http://www.geo.mtu.edu/upseis/images/rayleigh_animation.gif
6. "Speed of Sound through the Earth". Hypertextbook.com. Retrieved 2010-08-23.

7. Geographic.org. "Magnitude 8.0 SANTA CRUZ ISLANDS Earthquake Details". Gobal Earthquake Epicenters with Maps. Retrieved 2013-03-13.

8. Ohnaka, M. (2013). The Physics of Rock Failure and Earthquakes. Cambridge University Press. p. 148.ISBN 9781107355330

9. Spence, William; S. A. Sipkin; G. L. Choy (1989)."Measuring the Size of an Earthquake". United States Geological Survey. Retrieved 2006-11-03.

10. Geoscience Australia

11. Wyss, M. (1979). "Estimating expectable maximum magnitude of earthquakes from fault dimensions". Geology**7** (7):336–340. Bibcode:1979Geo.......336W. doi:10.1130/00 91-7613(1979)7<336:EMEMOE>2.0.CO;2.

12. Sibson R. H. (1982) "Fault Zone Models, Heat Flow, and the Depth Distribution of Earthquakes in the Continental Crust of the United States", Bulletin of the Seismological Society of America, Vol 72, No. 1, pp. 151–163

13. Sibson, R. H. (2002) "Geology of the crustal earthquake source" International handbook of earthquake and engineering seismology, Volume 1, Part 1, page 455, eds. W H K Lee, H Kanamori, P C Jennings, and C. Kisslinger, Academic Press, ISBN / ASIN: 0124406521

14. "Global Centroid Moment Tensor Catalog". Globalcmt.org. Retrieved 2011-07-24.

15. "Instrumental California Earthquake Catalog". WGCEP. Retrieved 2011-07-24.

16. Hjaltadóttir S., 2010, "Use of relatively located microearthquakes to map fault patterns and estimate the thickness of the brittle crust in Southwest Iceland"

17. "Reports and publications | Seismicity | Icelandic Meteorological office". En.vedur.is. Retrieved 2011-07-24.

18. Schorlemmer, D.; Wiemer, S.; Wyss, M. (2005). "Variations in earthquake-size distribution across different stress regimes". Nature 437 (7058):539–542. Bibcode: 2005 Natur.437..539S.doi:10.1038/nature04094. PMID 16177788.

19. Nettles, M.; Ekström, G.(May 2010). "Glacial Earthquakes in Greenland and Antarctica".Annual Review of Earth and Planetary Sciences 38 (1):467–491. Bibcode:2010AREPS..38..467N. doi:10.1146/annurev-earth-040809-152414. Avinash Kumar edit

20. Noson, Qamar, and Thorsen (1988).Washington State Earthquake Hazards: Washington State Department of Natural Resources. Washington Division of Geology and Earth Resources Information Circul

21. "M7.5 Northern Peru Earthquake of 26 September 2005" (PDF). National Earthquake Information Center. 17 October 2005. Retrieved 2008-08-01.

22. Greene II, H. W.; Burnley, P. C. (October 26, 1989). "A new self-organizing mechanism for deep-focus earthquakes". Nature 341 (6244): 733–737. Bibcode:1989Natur.341.. 733G. doi:10.1038/341733a0.

23. Foxworthy and Hill (1982). Volcanic Eruptions of 1980 at Mount St. Helens, The First 100 Days: USGS Professional Paper 1249.

24. Watson, John; Watson, Kathie (January 7, 1998)."Volcanoes and Earthquakes". United States Geological Survey. Retrieved May 9, 2009.

25. National Research Council (U.S.). Committee on the Science of Earthquakes (2003). "5. Earthquake Physics and Fault-System Science". Living on an Active Earth: Perspectives on Earthquake Science. Washington D.C.: National Academies Press. p. 418. ISBN 978-0-309-06562-7. Retrieved 8 July 2010.

26. Thomas, Amanda M.; Nadeau, Robert M.; Bürgmann, Roland (December 24, 2009). "Tremor-tide correlations and near-lithostatic pore pressure on the deep San Andreas fault". Nature **462** (7276):1048–51.Bibcode:2009Natur.462.1048T. doi:10.1038/nature 08654. PMID 20033046.

27. "Gezeitenkräfte: Sonne und Mond lassen Kalifornien erzittern" SPIEGEL online, 29.12.2009

28. Tamrazyan, Gurgen P. (1967). "Tide-forming forces and earthquakes". Icarus **7** (1–3): 59–65.Bibcode:1967Icar....7...59T. doi:10.1016/0019-1035(67)90047-4.

29. Tamrazyan, Gurgen P. (1968). "Principal regularities in the distribution of major earthquakes relative to solar and lunar tides and other cosmic forces". Icarus **9** (1–3): 574–92. Bibcode:1968Icar....9..574T. doi:10.1016/0019-1035(68)90050-X.

30. "What are Aftershocks, Foreshocks, and Earthquake Clusters?".

31. "Repeating Earthquakes". United States Geological Survey. January 29, 2009. Retrieved May 11, 2009.

32. "Earthquake Facts". United States Geological Survey. Retrieved 2010-04-25.

33. Pressler, Margaret Webb (14 April 2010). "More earthquakes than usual? Not really." KidsPost(Washington Post: Washington Post). pp. C10.

34. Madrigal, Alexis (4 June 2008). "Top 5 Ways to Cause a Man-Made Earthquake". Wired News (CondéNet). Retr

Resources

1. earthquake.usgs.gov/learn/kids/eqscience.php
2. earthsky.org/earth/the-science-behind-the-nepal-earthquake
3. https://nhmu.utah.edu/sites/default/.../All%20About%20Earthquakes.pdf
4. www.geo.mtu.edu/UPSeis/why.html
5. science.howstuffworks.com/nature/natural-disasters/earthquake.htm
6. www.wsj.com/.../how-the-nepal-earthquake-happened-like-clockwork-143...
7. https://en.wikipedia.org/wiki/Earthquake

5. Earthquake Effects

Abstract

Earthquakes are natural disasters of a generally unpredictable nature. In spite of considerable efforts made towards improving the understanding of these natural disasters and protecting built environment from their effects, earthquakes still cause huge human and economic losses; this is true both for highly industrialized and lesser developed countries.

The ground movements caused by earthquakes can have several types of damaging effects. Some of the major effects are:

- Ground shaking, i.e. back-and-forth motion of the ground, caused by the passing waves of vibration through the ground;
- Soil failures, such as liquefaction and landslides, caused by shaking;
- Surface fault ruptures, such as cracks, vertical shifts, general settlement of an area, landslides, etc.
- Tidal waves (tsunamis), i.e. large waves on the surface of bodies of water that can cause major damage to shoreline areas.
- Shaking and ground rupture are the main effects created by earthquakes, principally resulting in more or less severe damage to buildings and other rigid structures.

5.1. General Effects of Earthquakes

A general study of earthquakes includes: consideration of the nature of ground faults, the propagation of shock waves through the earth mass, the specific nature of recorded major quakes, etc. The basic concept of designing the buildings capable to resist earthquake effects is presented here.

5.1.1. *Animation and Visualization of an Earthquake* [1]

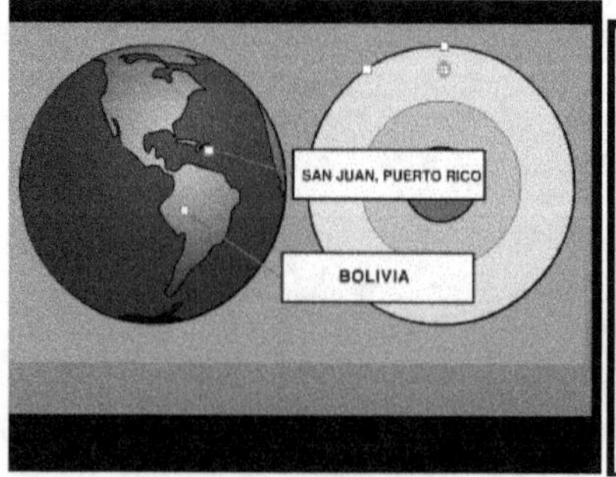

This animation is a way to visualize an earthquake as it travels outwards from its starting point through the earth (viewed from above at left; in cross-section at right). It is based on a real earthquake that took place in Bolivia in 1994. Seismic waves from the quake were recorded around the world, including at a recording station in San Juan, Puerto Rico.

Deep underground, a fault that has been storing strain suddenly ruptures, releasing the stresses like a spring uncoiling. This pumps seism waves into the surrounding rock. Earthquake!
The underground starting point of the earthquake is called the "focus."

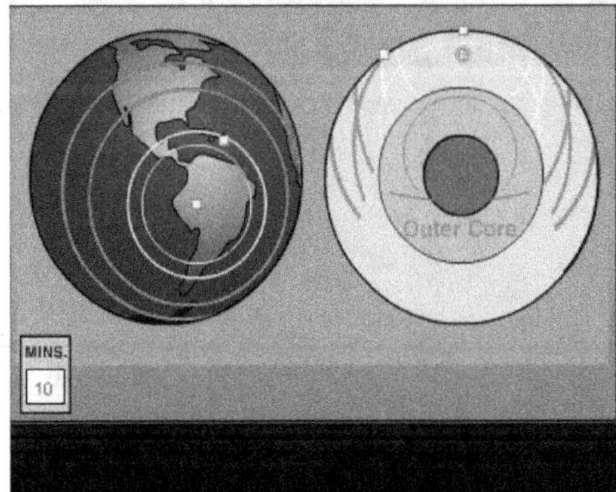

The waves travel outward in all directions from the quake's focus.
Two kinds of waves are shown here: P-waves, or primary waves, and S-waves, or secondary waves. (P-waves compress the rock through which they travel, while S-waves distort its shape or shear it.) Here, P-waves are red and S-waves are blue.

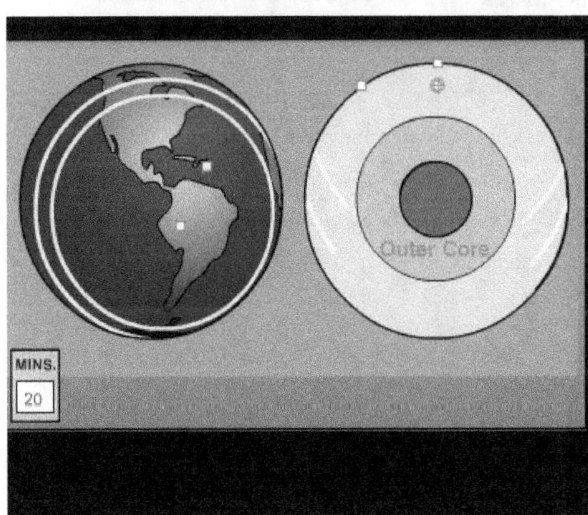

Primary waves travel faster. For example, the first P-waves reached San Juan 5 minutes after the rupture. The first S-waves did not arrive for another 4 minutes.
Another difference is that P-waves can travel through fluid, although they are slowed by it, but S-waves cannot. Therefore, only P-waves can travel through the molten core of the earth. The S-waves bounce off the core.

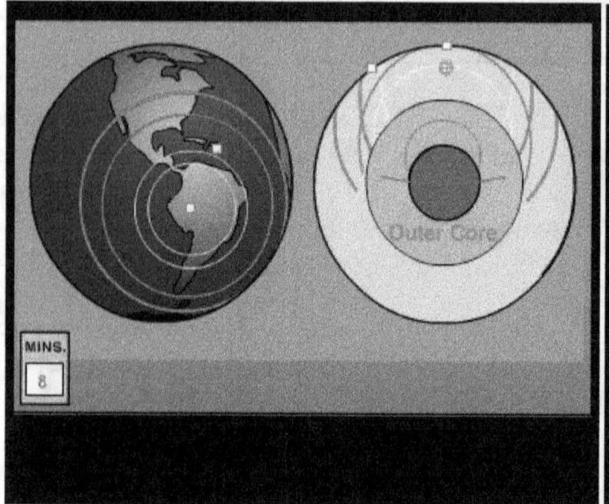

The ground movements caused by earthquakes can have several types of damaging effects. Some of the major effects are:

1. Ground shaking, i.e. back-and-forth motion of the ground, caused by the passing waves of vibration through the ground;

2. Soil failures, such as liquefaction and landslides, caused by shaking;

3. Surface fault ruptures, such as cracks, vertical shifts, general settlement of an area, landslides, etc.

4. Tidal waves (tsunamis), i.e. large waves on the surface of bodies of water that can cause major damage to shoreline areas.

5.1.2. Shaking and Ground Rupture

Shaking and ground rupture are the main effects created by earthquakes, principally resulting in more or less severe damage to buildings and other rigid structures. The severity of the local effects depends on the complex combination of the earthquake magnitude, the distance from the epicenter, and the local geological and geo-morphological conditions, which may amplify or reduce wave propagation.[2]The ground-shaking is measured by ground acceleration.

Specific local geological, geo-morphological, and geo-structural features can induce high levels of shaking on the ground surface even from low-intensity earthquakes. This effect is called site or local amplification. It is principally due to the transfer of the seismic motion from hard deep soils to soft superficial soils and to effects of seismic energy focalization owing to typical geometrical setting of the deposits.

Ground rupture is a visible breaking and displacement of the Earth's surface along the trace of the fault, which may be of the order of several meters in the case of major earthquakes. Ground rupture is a major risk for large engineering structures such as dams, bridges and nuclear power stations and requires careful mapping of existing faults to identify any which are likely to break the ground surface within the life of the structure.[3]

5.1.3. Landslides and Avalanches

Landslide: Earthquakes, along with severe storms, volcanic activity, coastal wave attack, and wildfires, can produce slope instability leading to landslides, a major geological hazard. Landslide danger may persist while emergency personnel are attempting rescue.[4]

5.1.4. Fires

Earthquakes can cause fires by damaging electrical power or gas lines. In the event of water mains rupturing and a loss of pressure, it may also become difficult to stop the spread of a fire once it has started.

5.1.5. Soil Liquefaction[5]

Soil liquefaction occurs when, because of the shaking, water-saturated granular material (such as sand) temporarily loses its strength and transforms from a solid to a liquid. Soil liquefaction may cause rigid structures, like buildings and bridges, to tilt or sink into the liquefied deposits. Liquefaction may occur when water-saturated sandy soils are subjected to earthquake ground shaking. When soil liquefies, it loses strength and behaves as a viscous liquid (like quicksand) rather than as a solid. This can cause buildings to sink into the ground or tilt, empty buried tanks to rise to the ground surface, slope failures, nearly level ground to shift laterally tens of feet (lateral spreading), surface subsidence, ground cracking, and sand blows. Liquefaction has caused significant property damage in many earthquakes around the world, and is a major hazard associated with earthquakes in Utah. The 1934 Hansel Valley and 1962 Cache Valley earthquakes caused liquefaction, and large prehistoric lateral spreads exist at many locations along the Wasatch Front. The valleys of the Wasatch Front are especially vulnerable to liquefaction because of susceptible soils, shallow ground water, and relatively high probability of moderate to large earthquakes.

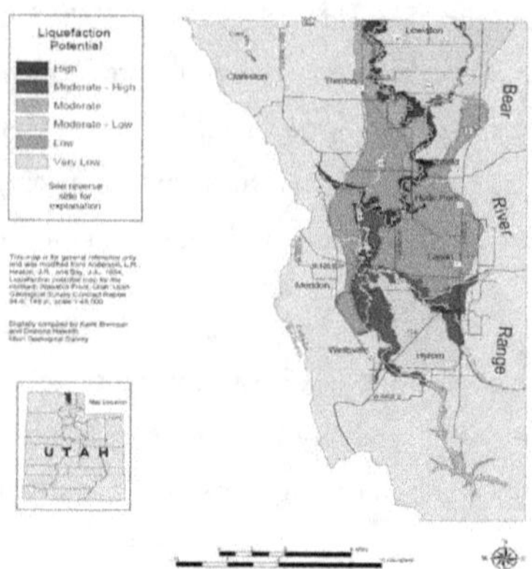

Image: Liquefaction Potential Map for Cache Valley-Cache Country, Utah-Utah Geological Survey (Public Information Series 79, August 2013)

Likely Occurrence of Liquefaction

Two conditions must exist for liquefaction to occur: (1) the soil must be susceptible to liquefaction (loose, water-saturated, sandy soil, typically between 0 and 30 feet below the ground surface) and (2) ground shaking must be strong enough to cause susceptible soils to liquefy. Northern, central, and southwestern Utah are the state's most seismically active areas. Identifying soils susceptible to liquefaction in these areas involves knowledge of the local geology and subsurface soil and water conditions. The most susceptible soils are generally along rivers, streams, and lake shorelines, as well as in some ancient river and lake deposits.

Determination of Liquefaction Potential

The liquefaction potential categories shown on this map depend on the probability of having an earthquake within a 100-year period that will be strong enough to cause liquefaction in those zones. High liquefaction potential means that there is a 50% probability of having an earthquake within a 100-year period that will be strong enough to cause liquefaction. Moderate means that the probability is between 10% and 50%, low between 5 and 10%, and very low less than 5%.

To determine the liquefaction potential and likelihood of property damage at a site, a site-specific geotechnical investigation by a qualified professional is needed. If a hazard exists, various hazard-reduction techniques are available, such as soil improvement or special foundation design. The cost of site investigations and/or mitigation measures should be balanced with an acceptable risk.

5.1.6. *Tsunami*

Tsunamis are long-wavelength, long-period sea waves produced by the sudden or abrupt movement of large volumes of water. In the open ocean the distance between wave crests can surpass 100 kilometers (62 mi), and the wave periods can vary from five minutes to one hour. Such tsunamis travel 600-800 kilometers per hour (373–497 miles per hour), depending on water depth. Large waves produced by an earthquake or a submarine landslide can overrun nearby coastal areas in a matter of minutes. Tsunamis can also travel thousands of kilometers across open ocean and wreak destruction on far shores hours after the earthquake that generated them.[6]

Ordinarily, sub-ducting earthquakes under magnitude 7.5 on the Richter scale do not cause tsunamis, although some instances of this have been recorded. Most destructive tsunamis are caused by earthquakes of magnitude 7.5 or more.[6]

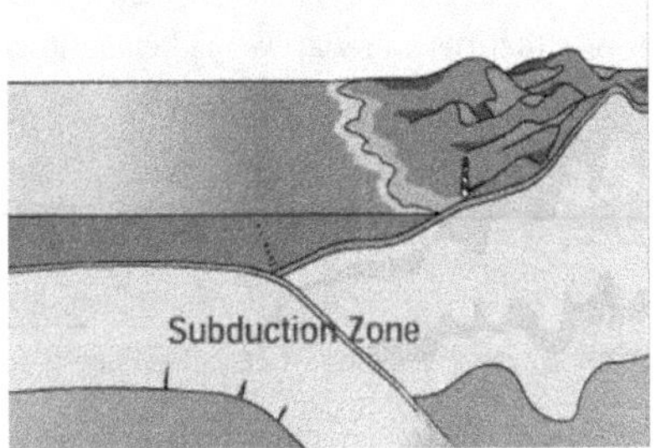

http://www.pbs.org/wnet/savageearth/animations/tsunami/main.html [7]

Tsunamis are usually caused by underwater earthquakes. These often occur offshore at sub-ducting zones (places where a tectonic plate that carries an ocean is gradually slipping under a continental plate).

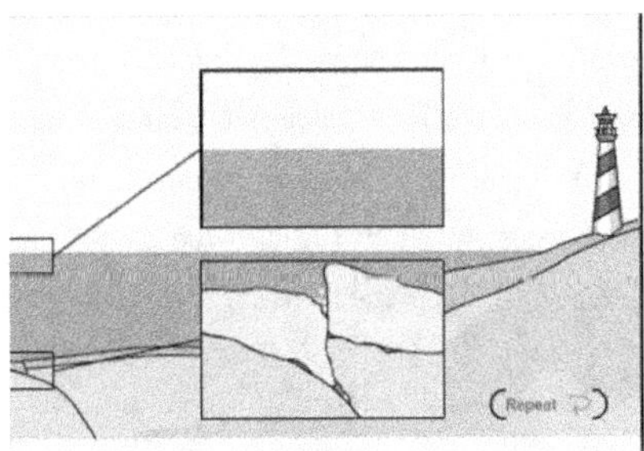

http://www.pbs.org/wnet/savageearth/animations/tsunami/middle2.html [8]

Part of the sea floor can snap upward abruptly, while other areas sink downward, when sections of the plates that have been locked together for a while move suddenly under the strain. In the instant after such an underwater earthquake, the shape of the sea surface mirrors the new contours of the sea floor--some areas of water are pushed upwards, and others sink.

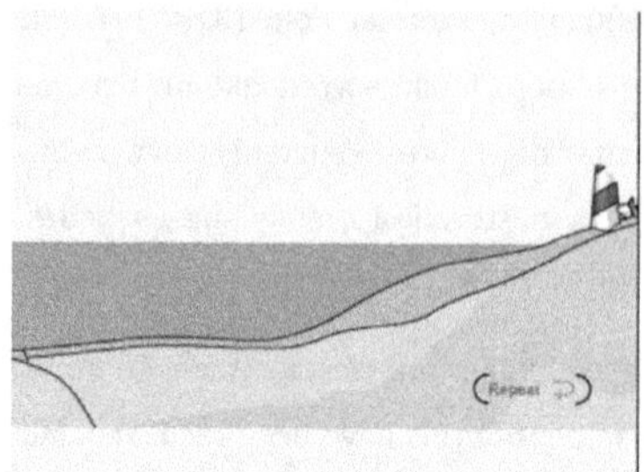

http://www.pbs.org/wnet/savageearth/animations/tsunami/middle3.html [9]

This starts a series of waves that rush outwards--the beginning of a tsunami. These waves travel very far and very fast (more than 500 miles/hour, or the speed of a jet airplane). At first, out at sea in deep water, the waves are very far apart--sometimes hundreds of miles--and their crests are not very high, perhaps only a few feet above the rest of the surface (although these crests are only the tips of vast masses of water in motion). Seen from a passing ship or low-flying plane, they would probably not even be noticeable. When a tsunami leaves deep water and approaches the shore, however, it slows down and its height grows. The wave crests also squeeze closer together. Depending on the shapes of the sea floor off the coast and of the coastline, a tsunami hitting the coast may appear as a series of towering walls of water that can level buildings.

A typical example is the tsunami of the 2004 Indian Ocean earthquake: The undersea mega-thrust earthquake was caused when the Indian Plate was sub ducted by the Burma Plate and triggered a series of devastating tsunamis along the coasts of most landmasses bordering the Indian Ocean, killing 230,000 people in 14 countries, and inundating coastal communities with waves up to 30 meters (100 ft) high[10] It was one of the deadliest natural disasters in recorded history. Indonesia was the hardest-hit country, followed by Sri Lanka, India, and Thailand. With a magnitude of M_w 9.1–9.3, it is the third-largest earthquake ever recorded on a seismograph.

5.1.7. *Volcanic Eruption* [11]

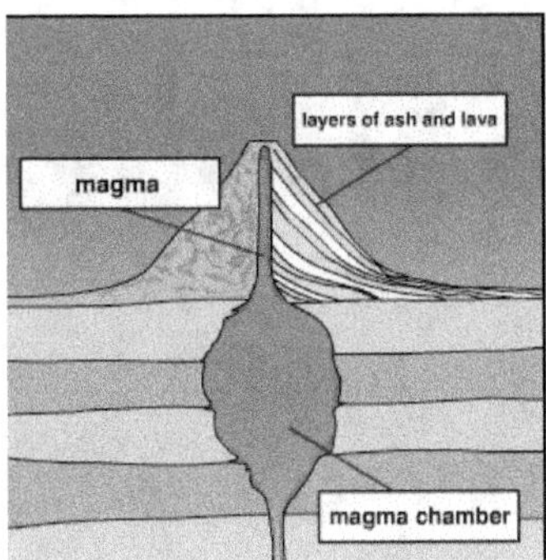

This is a type of volcano called a stratovolcano. It can be seen how this very tall volcano has been built up by layers of ash and lava from previous eruptions. The volcano is currently dormant. But it is about to re-awaken! Because it is lighter than solid rock, magma (molten rock) has been rising to form a huge pool, called a magma chamber, beneath the volcano.

The volcano will erupt when gases held within the magma burst out explosively, now that they are under less pressure than when the magma was deeper within the earth.

The explosive escape of the gases from the magma causes a violent eruption which literally "blows the top" off the volcano. Gas, blobs of magma (now called lava), chunks of rock, and fine debris called volcanic ash are blasted upwards. (The ash, because it is very light, may reach heights sufficient to endanger jet airplanes.)

Because the fragments of rock and ash (pyroclastics) are heavier than air and may not mix with enough air to remain high aloft, they may surge down the sides of the volcano as dangerous "pyroclastic flows." Within the flow, heavier particles sink and the lighter particles and gases are displaced upwards. The flow meanwhile surges along like a boiling cloud. Pyroclastic flows can travel at speeds greater than 100 miles per hour and can be as hot as the inside of a kiln. Such flows killed 29,000 people in the 1902 eruption of Mount Pelée, Martinique.

The Process that Leads to Volcanic Eruptions

- Swarms of small earthquakes precede eruptions
- The frequency of the earthquakes increases
- They reach a point of such rapid succession that they create a harmonic tremor
- The frequency of the harmonic tremor steadily rises
- The volcano 'screams' due to incredible pressure building up
- The tremor and the screaming stops
- The volcano erupts

5.1.8. Floods

A flood is an overflow of any amount of water that reaches land.[12] Floods occur usually when the volume of water within a body of water, such as a river or lake, exceeds the total capacity of the formation, and as a result some of the water flows or sits outside of the normal perimeter of the body. However, floods may be secondary effects of earthquakes, if dams are damaged. Earthquakes may cause landslips to dam rivers, which collapse and cause floods [13]

5.1.9. Human Impacts

An earthquake may cause injury and loss of life, road and bridge damage, general property damage, and collapse or destabilization (potentially leading to future collapse) of buildings. The aftermath may bring disease, lack of basic necessities, and higher insurance pre

5.1.10. Major Earthquakes

Earthquakes of magnitude 8 and greater fall under the category of major earthquakes

Earthquakes of magnitude 8.0 and greater since 1900 are shown in the Map below. The apparent 3D volumes of the bubbles are linearly proportional to their respective fatalities. [14]

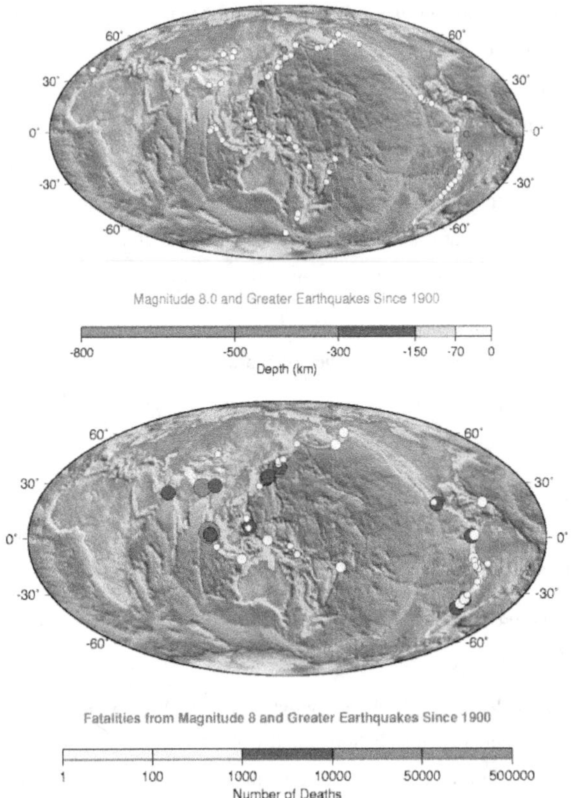

Earthquakes that caused the greatest loss of life, while powerful, were deadly because of their proximity to either heavily populated areas or the ocean, where earthquakes often create tsunamis that can devastate communities thousands of kilometers away.

5.2.   Building Reaction to Ground Motion

Sources

- Paper by Christopher Arnold; Resources)
- http://nisee.berkeley.edu/lessons/arnold.html
- http://commons.bcit.ca/civil/students/earthquakes/resources.htm

A couple of important terms related to building reaction to earthquakes, inertial forces and fundamental period of vibration are described here [15]:

5.2.1. *Inertial Forces*

When earthquake shaking occurs, a building gets thrown from side to side and/or up and down. That is, while the ground is violently moving from side to side, the building tends to stand at rest, similar to a passenger standing on a bus that accelerates quickly. Once the building starts moving, it tends to continue in the same direction, but by this time the ground is moving back in the opposite direction (as if the bus driver first accelerated quickly, then suddenly braked). Internal forces in a building caused by vibration of the building's mass during earthquake shaking are called **inertial forces**. The building's mass, size, and shape - its configuration - partially determine these forces and also partially determine how well they will be resisted.

Inertial forces are equal to the product of mass and acceleration as per the ***Newton's Second Law***

$$F = m \times a$$

Acceleration 'a' is the change of velocity (or speed in a certain direction) over time and is a function of the nature of the earthquake; mass 'm' is an attribute of the building. Since the forces are inertial, an increase in the mass generally results in an increase in the force.

Hence the immediate virtues of the use of light weight construction as a seismic design approach is called for.

The other detrimental aspect of mass, besides its role in increasing the lateral loads, is that failure of vertical elements such as columns and walls can occur by buckling when the mass pushing down due to gravity exerts its force on a member bent or moved out of plumb by the lateral forces. This phenomenon is known as the P-e, or P-Delta effect. The greater the vertical force, the greater the moment due to the product of the force, P, and the eccentricity, e (or Delta).

Although buildings generally have large vertical load-carrying reserves due to code gravity load requirements, this safety factor does not necessarily mitigate the P-e problem, which can induce bending in columns.

Earthquakes shake the ground in a variety of directions - including up and down components. Historically, codes generally treated these vertical earthquake forces lightly, although they may be two-thirds as great as the lateral earthquake forces, and "seismic design" and "design for lateral forces" are not really synonymous terms. It is vertical loads that almost always cause buildings to collapse in earthquakes; however, in earthquakes buildings generally fall down, not over. The lateral forces use up the strength of the structure by bending and shearing columns, beams, and walls, and then gravity pulls the weakened and distorted structure down.

It is important to note that the main difference between the nature of earthquake and wind loading is due to the fact that the earthquake ground motion induces internally generated inertial forces caused by vibration of the building's mass, whereas wind loading acts in the form of externally applied pressure.

5.2.2. *Fundamental Period of Vibration*

If one shook a flag pole with a heavy weight on top in the attempt to break it, one would quickly learn to synchronize one's pushes and pulls with the pole's natural tendency to vibrate back and forth at a certain rate - its fundamental period. If it tends to swing back and forth one complete cycle once a second when "plucked" and allowed to vibrate, it has a fundamental period of one second. If we can predict approximately the rate at which the ground will shake, which is similar to controlling the rate at which one shakes the base of the pole by hand, we could adjust the rate at which the pole will naturally vibrate so that the two either will or will not coincide. If they coincide, then the dimensions of the swing will start to increase, the pole will be said to resonate, and the loads on it will increase.

Ground motion will impart vibrations to a building of a similar nature to our shaking of the flag pole. The fundamental periods of structures may range from about 0.05 second for a well anchored piece of equipment, 0.1 second for a one story simple bent or frame, 0.5 second for a low structure up to about 4 stories, and between 1-2 seconds for a tall building from 10 20 stories. A water tank on an offshore drilling rig will be between 2.5 and 6 seconds, and a large suspension bridge may have a period of about 6 seconds.

An illustrative animation of vibrations of a braced frame steel structure is presented below. [15] This animation file has been developed using **SAP2000**, a structural analysis software package.

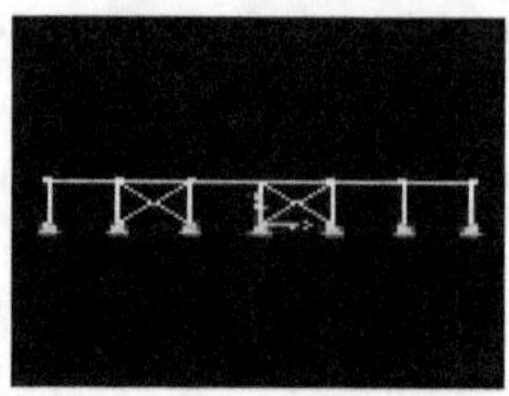

Image: Animation of Vibrations of a Braced Frame Steel Structure [15]

Natural periods of soil are usually in the range of 0.5 to 1 second, so that it is possible for the building and ground to have the same fundamental period and therefore there is a high probability for the building to approach a state of partial resonance (quasi-resonance). Hence in developing a design strategy for a building, it is desirable to estimate the fundamental periods both of the building and of the site so that a comparison can be made to see if the probability of quasi-resonance exists. If the initial study shows this to be the case, then it would be advisable to change the resonance characteristics of the building (for the site characteristics are fixed).

5.2.3. *Earthquake Effects on Building* [16]

There are certain key characteristics of earthquake ground shaking that affect the seismic performance of buildings. In addition, there are also building characteristics that, together with those of the ground, determine the building's seismic performance (how much damage the building will suffer). These characteristics are common to all buildings, both new and existing, and all locations. The building response to earthquake shaking occurs over the time of a few seconds. During this time, the several types of seismic waves are combining to shake the building in ways that are quite different for each earthquake. In addition, as the result of variations in fault slippage, differing rock through which the waves pass, and the different geological nature of each site, the resultant shaking at each site is different. The characteristics of each building are different, whether in size, configuration, material and structural system, method of analysis, age, or quality of construction-each of these characteristics affects the building response. In spite of the complexity of the interactions between the building and the ground during the few seconds of shaking there is broad understanding of how different building types will perform under different shaking conditions? This understanding comes from extensive observation of buildings in earthquakes all over the world, together with analytical and experimental research at many universities and research centers. Understanding the ground and building characteristics is essential to give designers a "feel" for how their building will react to shaking, which is necessary to guide the conceptual design of their building. There should be a focus on certain architectural characteristics that influence seismic performance - either in a positive or negative way.

5.3. Building Diagnostics for Earthquake Resistance [17]

Shinko Plan-tech performs earthquake-resist diagnostics of plant buildings and equipment mounts based on their long-standing experience and broad expertise. Earthquake resist diagnostics are performed on existing structures and plant foundations, including the review of potential equipment modifications.

Advantages

- Earthquake-resist diagnostics are performed in accordance with guidelines established by the Law on the Promotion of Seismic Retrofitting of Buildings (1995). (Earthquake-resist targets are in line with the current Building Standards Law.)
- Capable of performing a diagnostics and assessment of joint deficiencies seen in most aging structures to enhance earthquake-resist.
- Assessments are made based on strength and stability, while diagnoses are performed on ultimate strength. This ensures structural integrity and safety during large earthquakes with the smallest amount of reinforcement possible.
- The calculation model is based on existing design drawings. When sufficient materials are unavailable, a separate on-site survey is performed for confirmation.
- When reinforcement is necessary, diagnostics results are used to provide the most appropriate solution, based on a comprehensive overview of the facilities and equipment.

Services

- Earthquake-resist diagnostics for plant buildings, general buildings and equipment mounts (structures)
- Planning, design and construction of building reinforcement and equipment modification
- Various survey planning and construction

Basic Business Outline

- Document search / verification
- On-site inspection (comparing design drawing with actual conditions, rough size measurement, visual inspection of aging)
- Diagnostic standard: voluntary diagnostics based on the standards set by the Japan Building Disaster Prevention Association
- Report: summarizes the inspection result and includes general findings
- Reinforcement design proposal compiled for implementation as needed

Diagnostics Flow for Steel Structures

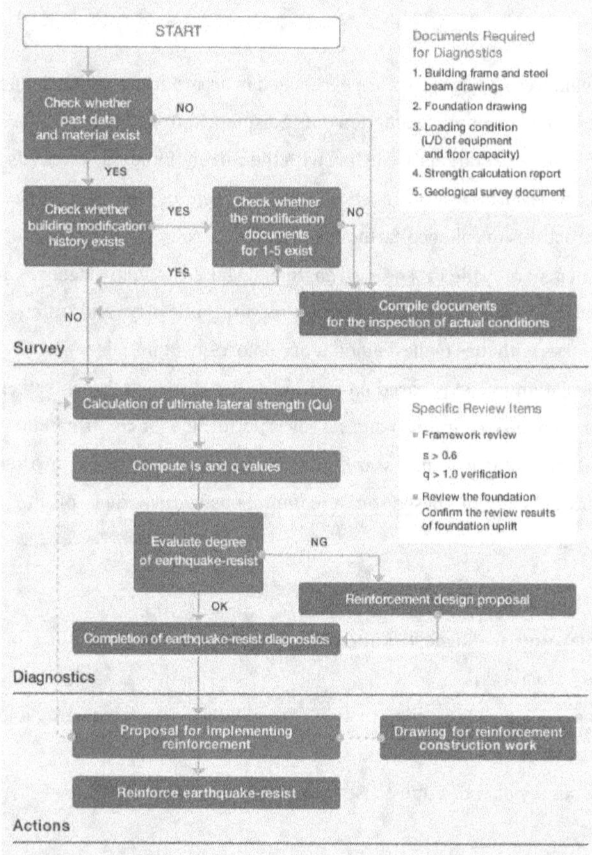

Precaution Statement

- This diagnostics service is for clients who are concerned about the safety of buildings and plant foundations in the event of a large earthquake, and is not an official review under applicable laws or regulations that require approval for facility modifications.
- The earthquake-resist target is for the structure to tolerate moderate structural changes and prevent collapse.
- This diagnostics is compiled as a voluntary survey and is not an applicable third-party assessment for certification purposes.
- When performing an estimate of diagnostics costs, Shinko Plan-tech must have access to client specifications, such as existing design drawings, total floor area and number of levels, etc.
- We require access to design drawings and structural calculation sheets, and will perform the drafting of current condition drawings, building weight and axial strength calculations separately.

Forced Vibrations Resonance [18]

Resonance

The phenomenon of driving a system with a frequency equal to its natural frequency is called resonance. As the driving frequency gets progressively higher than the resonant or natural frequency, the amplitude of the oscillations becomes smaller until the oscillations nearly disappear.

Natural Frequency

The frequency at which a system vibrates on its own is called natural frequency. For a spring (with spring constant **k**) with an object of mass **m** attached, the natural frequency is given as $f_n = \frac{1}{2\pi}\sqrt{\frac{k}{m}}$.

Damping

The reduction in the magnitude of oscillations by the dissipation of energy

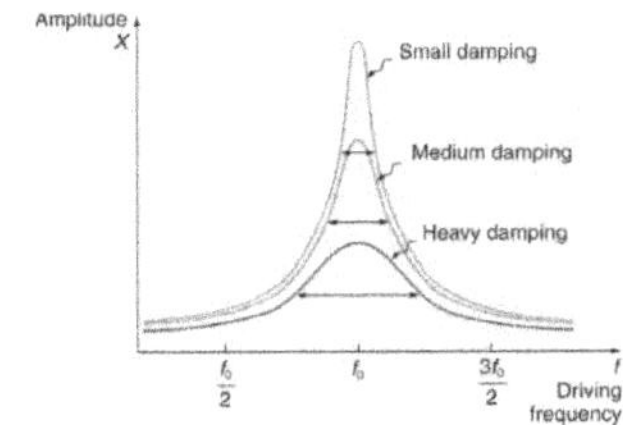

The amplitude of a harmonic oscillator is a function of the frequency of the driving force. The curves represent the same oscillator with the same natural frequency but with different amounts of damping. Resonance occurs when the driving frequency equals the natural frequency, and the greatest response is for the least amount of damping. The narrowest response is also for the least amount of damping

Seismic Isolator for Buildings [19]

Seismic isolation structure protects lives and property from earthquake, Nowadays, "seismic isolation structure" is regarded as a technology to prevent damage or collapse of buildings from earthquake. Seismic isolation bearing can prevent damage to buildings; it is also effective to ensure the safety inside the building.

When the large earthquake energy is transmitted to buildings, the buildings may suffer from damage or collapse. How to deal with earthquakes, methods of construction or structure has been devised and measured by earthquake prone country, Japan. The buildings have 3 types of structure, seismic isolation, seismic vibration control and seismic resistant structures. Illustration below (a, b, c) describes the differences of each other:

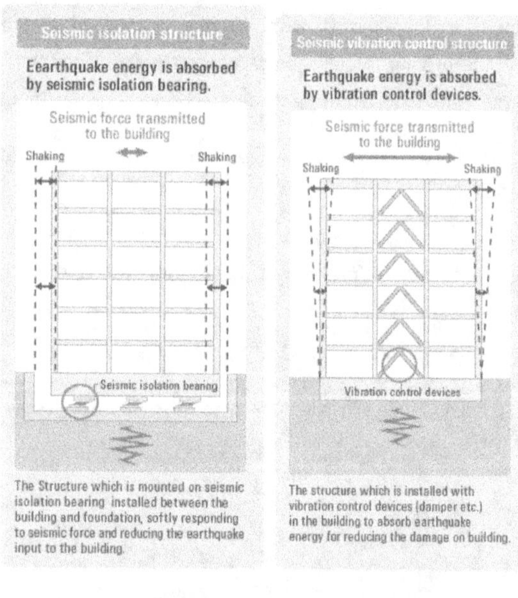

(a) (b)

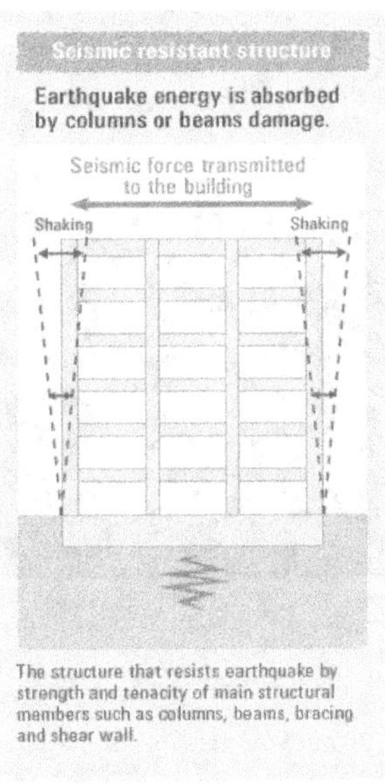

(c)

Performance of Seismic Isolation Bearing

Seismic isolated building is mounted on seismic isolation bearings. While supporting the weight of building, seismic isolation bearings can perform large horizontal movement to protect the building from earthquake. In developing a design strategy for a building, it is desirable to estimate the fundamental period of vibration for both of the building as well as the site for comparison. In case there exists the probability for quasi-resonance then it would be advisable to change the resonance characteristics of the building (because the site characteristics are fixed).

5.4.   Forced Vibrations [20]

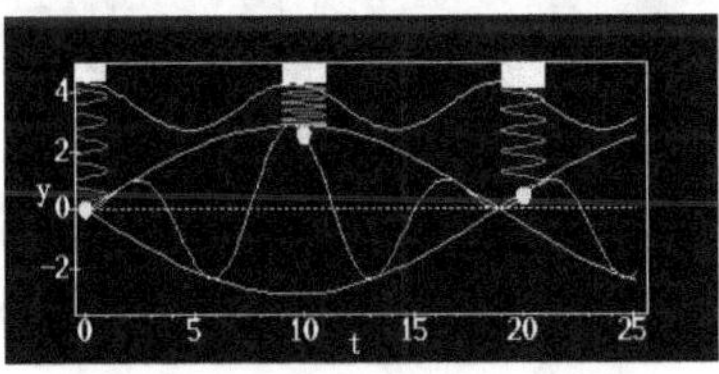

Image: Undamped Forced Vibrations

bts.gif: www.math.ubc.ca500 × 180Search by image

5.4.1. *Undamped Forced Vibrations*

Here is the case $\omega_0 = 1, \omega = \frac{2}{3}$ the external force coming from moving the support of the spring up and down with amplitude 0.8.

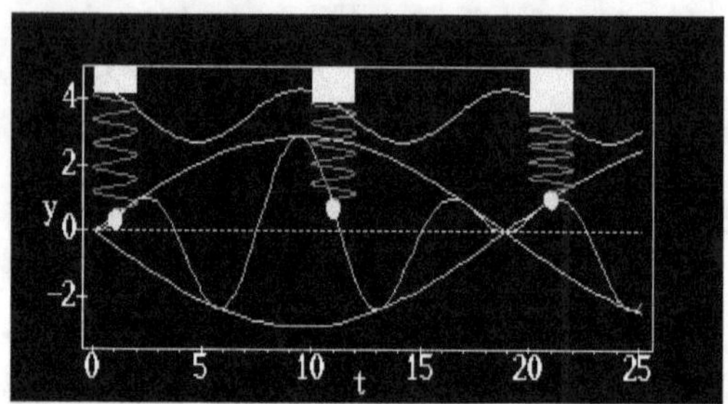

The effect is that of an angular frequency that is the average of ω_0 and ω, with an amplitude that varies because of the low-frequency factor $\sin\left(\frac{\omega - \omega_0}{2}t\right)$. This phenomenon is known as ``beats'': in an audio signal you hear the signal grow louder and softer, dying out when$(\omega - \omega_0)^{t/2}$ is a multiple of π and sounding loudest when it is an odd multiple of $\frac{\pi}{2}$.

5.4.2. *Resonance*

It describes an oscillation with amplitude that grows steadily. This phenomenon is called resonance of course, in a real physical system the amplitude can't grow forever, and eventually the differential equation will not be valid. In many cases the reason the equation is no longer valid is that something breaks. This is why resonance is generally a bad thing.

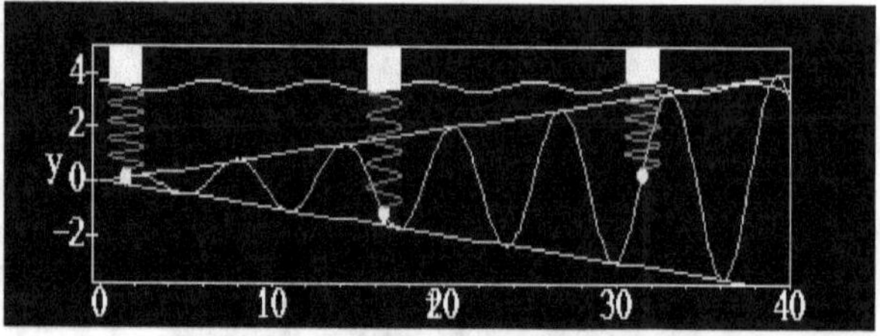

The picture shows spring-mass system, starting from rest, with the support moving up and down at the resonant frequency with amplitude 0.2. Eventually the mass collides with the support of the spring.

5.4.3. Damped Forced Vibrations

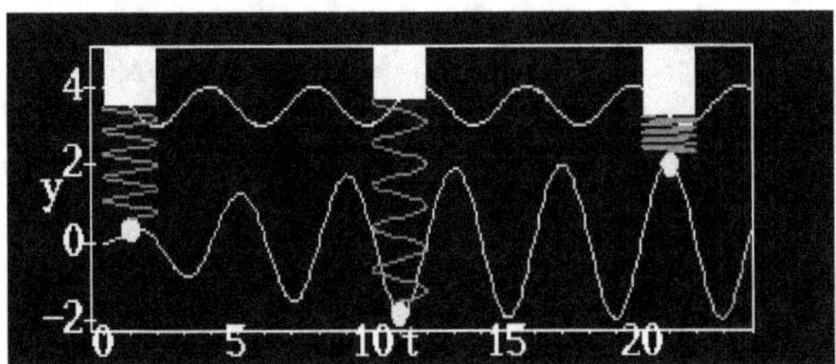

For any given ω and ω_0 the amplitude of the steady state solution will be less than the amplitude of the corresponding solution of the undamped system, and as the damping constant increases the amplitude will decrease. Under otherwise identical conditions, if ω_0 is allowed to vary, then the maximum amplitude will occur when $\omega_0 = \omega$.

Thus there are three effects of damping on forced vibrations:

1. It causes the solutions of the homogeneous equation to die away, so that after a long enough time all you see is the steady state solution.
2. It shifts the phase of the response.
3. It reduces the amplitude of the response

References

1. http://www.pbs.org/wnet/savageearth/animations/earthquakes/main.html
2. "On Shaky Ground, Association of Bay Area Governments, San Francisco, reports 1995, 1998 (updated 2003)". Abag.ca.gov. Retrieved 2010-08-23.
3. "Guidelines for evaluating the hazard of surface fault rupture, California Geological Survey" (PDF). California Department of Conservation. 2002.
4. "Natural Hazards-Landslides". United States Geological Survey. Retrieved 2008-09-15.
5. https://nhmu.utah.edu/sites/default/.../All%20About%20Earthquakes.pdf
6. .[a][b] Noson, Qamar, and Thorsen (1988). Washington Division of Geology and Earth Resources Information Circular 85. Washington State Earthquake Hazards.
7. http://www.pbs.org/wnet/savageearth/animations/tsunami/main.html
8. http://www.pbs.org/wnet/savageearth/animations/tsunami/middle2.html
9. http://www.pbs.org/wnet/savageearth/animations/tsunami/middle3.html
10. https://en.wikipedia.org/wiki/2004_Indian_Ocean_earthquake_and_tsunami
11. http://www.pbs.org/wnet/savageearth/animations/volcanoes/main.html
12. MSN Encarta Dictionary. Flood. Retrieved on 2006-12-28. Archived 2009-10-31.

13. "Notes on Historical Earthquakes". British Geological Survey. Retrieved 2008-09-15.

14. USGS: Magnitude 8 and Greater Earthquakes Since 1900

15. http://commons.bcit.ca/civil/students/earthquakes/unit2_02.htm

16. https://www.s-plantech.co.jp/e/service/engineering/earthquake01.html [20]

17. http://www.fema.gov/media-library-data/20130726-1556-20490-0102/fema454_chapter4.pdf [21]

18. https://www.boundless.com/physics/textbooks/boundless-physics-textbook/sound-16/further-topics-132/forced-vibrations-and-resonance-473-1966/ [22]

19. http://www.bridgestone.com/products/diversified/antiseismic_rubber/movie/ [23]

20. www.math.ubc.ca/~israel/m215/forced/forced.html

6. Earthquakes and Tsunamis in India- 2015 Nepal Earthquakes in Particular

The Indian subcontinent has a history of earthquakes. The reason for the high frequency and intensity of earthquakes is the Indian plate driving into Asia at a rate of approximately 49 mm/year. As a result one should constantly remain on alert. If the main centre exists in Himalaya range then it will damage maximum part of northern India. It is to be noted that the Himalayas being a fairly young mountain range is undergoing constant geological changes.

India, due to its, physiographic and climatic conditions is one of the most disaster prone areas of the world. It is vulnerable to windstorms from both the Arabian Sea and Bay of Bengal. There are active crustal movements in the Himalaya leading to earthquakes. About 58.7 % of the total land mass is prone to earthquake of moderate to very high intensity. Further, India has increasingly become vulnerable to Tsunamis since the 2004 Indian Ocean Tsunami. It is important to mention that India has a coastline running 7600 km long; as a result it is repeatedly threatened by cyclones.

India is also very prone to earthquakes as well. The major reason for the high frequency and intensity of the earthquakes is that the Indian plate is driving into Asia at a rate of approximately 47 mm/year. As per the Geographical statistics, almost 54% of the land in India is vulnerable to earthquakes.

According to the estimates shown by a World Bank and United Nations report; around 200 million city dwellers in India will be exposed to storms and earthquakes by 2050.

Earthquake Prone Cities in India

The Indian subcontinent has a history of devastating earthquakes. An earthquake of 7.2 magnitude struck Tajikistan and the tremors were felt in India and Pakistan.

The epicenter of the earthquake was Tajikistan, measuring 7.2 on the Richter Scale and was recorded at 25 km.

The latest version of the seismic zoning map of India assigns four levels of seismicity for India in terms of zone factors, which means India is divided in to 4 seismic zones:

1. Zone 2
2. Zone 3
3. Zone 4
4. Zone 5

Zone 5 is highly prone to the earthquake with the highest level of seismicity whereas Zone 2 is associated with the lowest level of seismicity. So, the Zones - marked two to five -indicate areas most likely to experience tremors with zone five being the most vulnerable.

Indian cities, ranging from the metros to the smaller cities - all at least once have been shaken up due to earthquakes which usually range from medium to high intensity on the Richter scale.

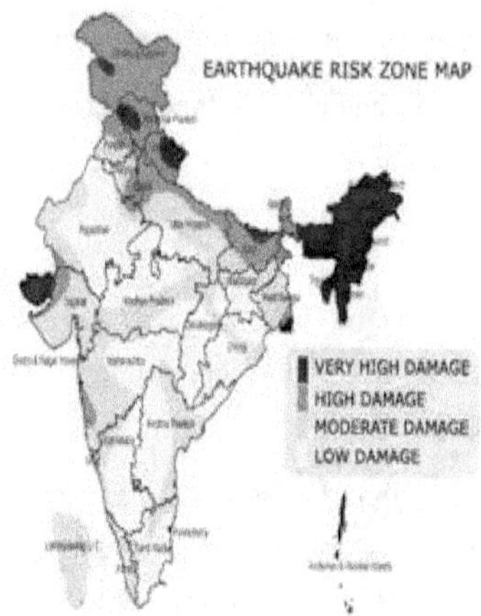

Earthquake Prone Cities in India

6.1.   Reason behind the Nepal Earthquake

In order to illustrate devastations caused due to an earthquake, a typical earthquake that occurred in Nepal on 25 April 2015 has been described in detail.

Experts explain the science behind Nepal's deadly earthquake based on continental collision. As the Indian sub continent pushes against Eurasia, pressure is released in the form of earthquake. The constant crashing of the two plates forms the Himalayan mountain range.

A little before noon on 25 April (Saturday) 2015 in Nepal, a chunk of rock about 9 miles below the earth's surface shifted, unleashing a shock wave—described as being as powerful as the explosion of more than 20 thermonuclear weapons—that ripped through the Katmandu Valley.

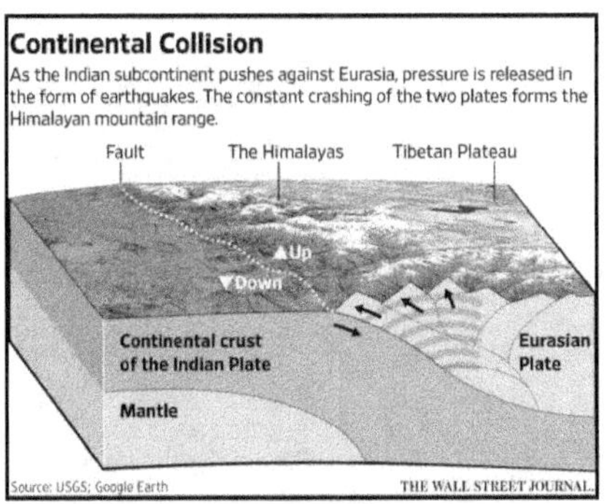

Source:http://www.wsj.com/articles/how-the-nepal-earthquake-happened-like-clockwork-1430044358 (Science behind the Nepal's Earthquake--Sean Mclain and Shirley s. Wang)

In geological terms, the tremor occurred like clock-work, 81 years after the region's last earthquake of such a magnitude, in 1934.Records dating to 1255 indicate the region—known as the Indus-Yarlung suture zone—experiences a magnitude-8 earthquake approximately every 75 years, according to a report by Nepal's National Society for Earthquake Technology.

The reason is the regular movement of the fault line that runs along Nepal's southern border, where the Indian subcontinent collided with the Eurasia plate 40 million to 50 million years ago.

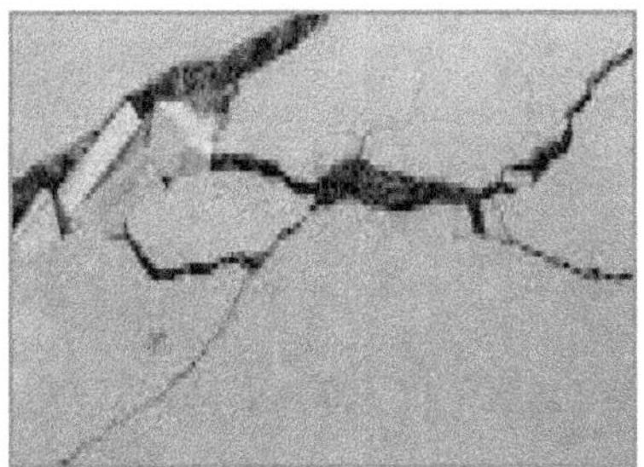

Photo: Agency France-Press/Getty Images

6.2. Damage from Saturday's Earthquake in Katmandu (Nepal)

"The collision between India and Eurasia is a showcase for geology," said Lung S. Chan, a geophysicist at the University of Hong Kong. The so-called India plate is pushing its way north toward Asia at a rate of about 5 centimeters, or 2 inches, a year, he said. "Geologically speaking, that's very fast."

As the plates push against each other, friction generates stress and energy that builds until the crust ruptures, said Dr. Chan, who compared the quake to a thermonuclear weapons explosion. In the case of Saturday's quake, the plate jumped forward about 2 meters, or 6.5 feet, said Hongfeng Yang, an earthquake expert at the Chinese University of Hong Kong.

Saturday's quake was also relatively shallow, according to the U.S. Geological Survey. Such quakes tend to cause more damage and more aftershocks than those that occur deeper below the earth's surface.

After an earthquake, the plates resume moving and the clock resets. *'Earthquakes dissipate energy, like lifting the lid off a pot of boiling water. But it builds back up after you put the lid back on.'*—Lung S. Chan, geophysicist at the University of Hong Kong

Nepal is prone to destructive earthquakes, not only because of the massive forces involved in the tectonic collision, but also because of the type of fault line the country sits on. Normal faults create space when the ground cracks and separates. Nepal lies on a so-called thrust fault, where one tectonic plate forces itself on top of another.

The most visible result of this is the Himalayan mountain range. The fault runs along the 1,400-mile range, and the constant collision of the India and Eurasia plates pushes up the height of the peaks by about a centimeter each year.

Despite the seeming regularity of severe earthquakes in Nepal, it isn't possible to predict when one will happen. Historic records and modern measurements of tectonic plate movement show that if the pressure builds in the region in a way that is "generally consistent and homogenous," the region should expect a severe earthquake every four to five decades, Dr. Yang said.

The complexity of the forces applying pressure at the fault means scientists are incapable of predicting more than an average number of earthquakes that a region will experience in a century, experts say.

Still, earthquakes in Nepal are more predictable than most, because of the regular movement of the plates. Scientists aren't sure why this is.

The earth's tectonic plates are constantly in motion. Some faults release built-up stress in the form of earthquakes. Others release that energy quietly. "Some areas, like Nepal, release energy as a large earthquake, once in a while," said Dr. Chan. "These regions all have different natures for reasons geologists don't really know."

6.3. Facts and Figures: 7.8M_w(M_w: Moment Magnitude Scale)

Date: 25April2015

Origin time	11:56:26 NST[1]
Magnitude	7.8M_w[1] or 8.1 M_s[2]
Depth	15.0 km (9.3 mi)[1]
Epicenter	28.147°N 84.708°E[1]
Type	Thrust[1]
Areas affected	• Nepal • India • China • Bangladesh
Total damage	≈$5 billion (about 25% of GDP)[3]
Max. intensity	IX (*Violent*)[1]
Aftershocks	7.3M_w on 12 May at 12:51[4] 6.7M_w on 26 April at 12:54[5] No. of aftershocks(>=4ML)=329 (as of 24 June 2015)[6]
Casualties	8,857 dead in Nepal (officially) and 9,018 in total[7][8] 21,952 injured (officially)[7]

Note: 7.8M_w = (7.8 Moment magnitude scale) , 8.1M_s= (8.1 Surface wave magnitude)

The **April 2015 Nepal earthquake** (also known as the **Gorkha earthquake**)[6][9] killed more than 9,000 people and injured more than 23,000. It occurred at 11:56 NST on 25 April, with a magnitude of 7.8M_w[1] or 8.1M_s[2]and a maximum Mercalli Intensity of IX (*Violent*). Its epicenter was east of the district of Lamjung, and its hypocenter was at a depth of approximately 15 km (9.3 mi).[1] It was the worst natural disaster to strike Nepal since the 1934 Nepal–Bihar earthquake.[10][11][12]

The earthquake triggered an avalanche on Mount Everest, killing at least 19,[13] making April 25, 2015 the deadliest day on the mountain in history.[14]The earthquake triggered another huge avalanche in the Langtang valley, where 250 people were reported missing.[15][16]

Hundreds of thousands of people were made homeless with entire villages flattened,[15][17][18] across many districts of the country. Centuries-old buildings were destroyed at UNESCO World Heritage sites in the Kathmandu Valley, including some at the Kathmandu Durbar Square, the Patan Durbar Squar, the Bhaktapur Durbar Square, the Changu Narayan Temple and the Swayambhunath Stupa. Geophysicists and other experts had warned for decades that Nepal was vulnerable to a deadly earthquake, particularly because of its geology, urbanization, and architecture.[19][20]

Continued aftershocks occurred throughout Nepal within 15–20 minute intervals, with one shock reaching a magnitude of 6.7 on 26 April at 12:54:08NST.[5] The country also had a continued risk of landslides.[21]

A major aftershock occurred on 12 May 2015 at 12:51 NST with a moment magnitude (M_w) of 7.3.[22] The epicenter was near the Chinese border between the capital of Kathmandu and Mt. Everest.[23] More than 200 people were killed and more than 2,500 were injured by this aftershock.[24]

6.3.1. The Nepal Main Earthquake and Aftershocks Map

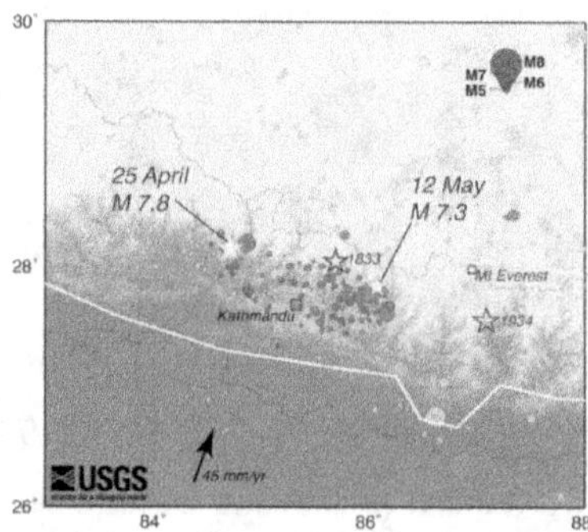

Map of the earthquake and aftershocks at 12 May, showing location of major historical earthquakes.

The earthquake occurred on 25 April 2015 at 11:56 a.m. NST (06:11:26 UTC) at a depth of approximately 15 km (9.3 mi) (which is considered shallow and therefore more damaging than quakes that originate deeper in the ground),[25] with its epicentre approximately 34 km (21 mi) east-southeast of Lamjung, Nepal, lasting approximately fifty seconds.[26] The earthquake was initially reported as 7.5 M_w by the United States Geological Survey (USGS) before it was quickly upgraded to 7.8 M_w. TheChina Earthquake Networks Center (CENC) reported the earthquake's magnitude to be 8.1 M_s. The India Meteorological Department (IMD) said two powerful quakes were registered in Nepal at 06:11 UTC and 06:45 UTC. The first quake measured 7.8 M_w and its epicenter was identified at a distance of 80 km to the northwest of Kathmandu, the capital of Nepal.Bharatpur was the nearest major city to the main earthquake, 53 km (33 mi) from the epicenter. The second earthquake was somewhat less powerful at 6.6 M_w. It occurred 65 km (40 mi) east of Kathmandu and its seismic focus lay at a depth of 10 km (6.2 mi) below the earth's surface. Over thirty-eight aftershocks of magnitude 4.5 M_w or greater occurred in the day following the initial earthquake, including the one of magnitude 6.6 M_w.[27]

According to the USGS, the temblor was caused by a sudden thrust, or release of built-up stress, along the major fault line where the Indian Plate, carrying India, is slowly diving underneath the Eurasian Plate, carrying much of Europe and Asia.[25]Kathmandu, situated on a block of crust approximately 120 km (74 miles) wide and 60 km (37 miles) long, reportedly shifted 3 m (10 ft) to the south in just 30 seconds.[28]

The risk of a large earthquake was well known beforehand. In 2013, in an interview with seismologist Vinod Kumar Gaur,*The Hindu* quoted him as saying, "Calculations show that there is sufficient accumulated energy [in the Main Frontal Thrust], now to produce an 8 magnitude earthquake. I cannot say when. It may not happen tomorrow, but it could possibly happen sometime this century, or wait longer to produce a much larger one."[29] According to Brian Tucker, founder of a nonprofit organization devoted to reducing casualties from natural disasters, some government officials had expressed confidence that such an earthquake would not occur again. Tucker recounted a conversation he had had with a government official in the 1990s who said, "We don't have to worry about earthquakes anymore, because we already had an earthquake"; the previous earthquake to which he referred occurred in 1934. [30]

6.3.2. Geology

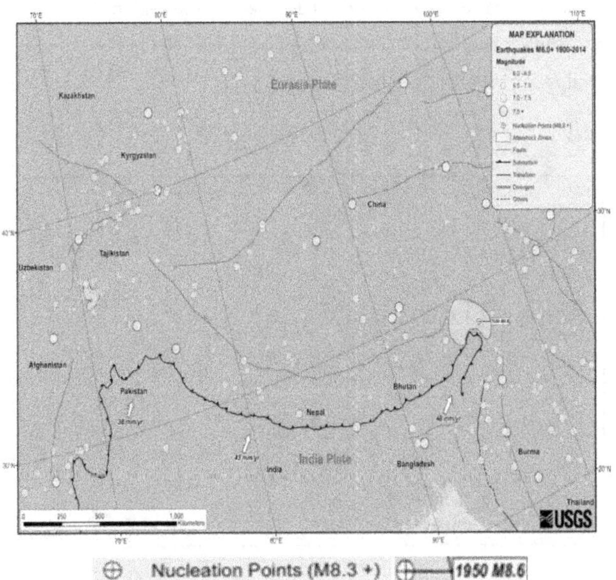

M6+ Himalayan region earthquakes, 1900–2014

Nepal lies towards the southern limit of the diffuse collision boundary where the Indian Plate under thrusts the Eurasian Plate,[31][32] occupying the central sector of the Himalayan arc, nearly one-third of the 2,400 km (1,500 mi) long Himalayas. Geologically, the Nepal Himalayas are sub-divided into five tectonic zones from north to south, east to west and almost parallel to sub-parallel.[33] These five distinct Morphed-Geotectonic zones are: (1) Terai Plain, (2) Sub Himalaya (SivalikRange), (3) Lesser Himalaya (Mahabharat Range and mid valleys), (4) Higher Himalaya, and (5) Inner Himalaya (Tibetan Tethys).[34] Each of these zones is clearly identified by their morphological, geological, and tectonic features.[34]

The convergence rate between the plates in central Nepal is about 45 mm (1.8 in) per year. The location, magnitude, and focal mechanism of the earthquake suggest that it was caused by a slip along the Main Frontal Thrust.[1][35]

The earthquake's effects were amplified in Kathmandu as it sits on the Kathmandu Basin, which contains up to 600 m (2,000 ft) of sedimentary rocks, representing the infilling of a lake.[36]

Based on a study published in 2014, of the Main Frontal Thrust, on average a great earthquake occurs every 750 ± 140 and 870 ± 350 years in the east Nepal region.[37] A study from 2015 found a 700-year delay between earthquakes in the region. The study also suggests that because of tectonic stress buildup, the earthquake from 1934 in Nepal and the 2015 quake are connected, following a historic earthquake pattern.[38]

6.3.3. Intensity

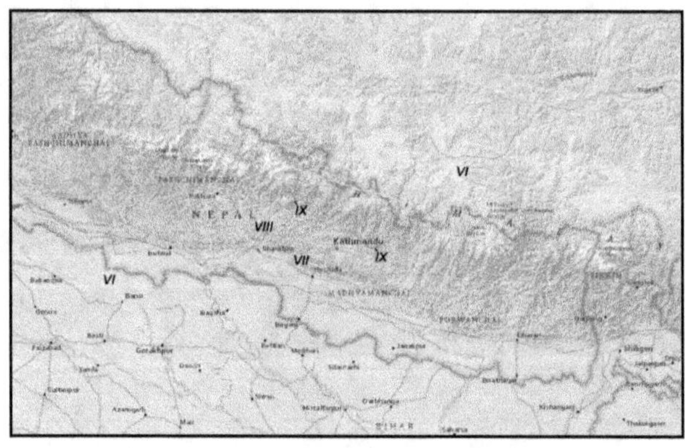

Isoseismic Map for the Gorkha Earthquake Annotated with Values on the Mercalli Scale

According to "Did You Feel It?" (DYFI?) responses on the USGS website, the intensity in Kathmandu was IX (*Violent*).[1] Tremors were felt in the neighboring Indian states of Bihar, Uttar Pradesh, Assam, West Bengal, Sikkim, Jharkhand, Uttarakhand, Gujarat [39] [better source needed] in the National capital region around New Delhi[40] and as far south as Karnataka.[41] Many buildings were brought down in Bihar. Minor cracks in the walls of houses were reported in Odisha. Minor quakes were registered as far as Kochi in the southern state of Kerala. The intensity in Patna was V (*Moderate*).[42] The intensity was IV (*Light*) in Dhaka, Bangladesh.[1] The earthquake was also experienced across southwestern China, ranging from the Tibet Autonomous Region to Chengdu, which is 1,900 km (1,200 mi) away from the epicenter.[43] Tremors were felt in Pakistan[44] and Bhutan.[1]

6.3.4. Aftershocks

List of Aftershocks after the 2015 Nepal Earthquake

A major aftershock of magnitude 6.7 M_w occurred on 26 April 2015 in the same region at 12:55 NST (07:09 UTC), with an epicenter located about 17 km (11 mi) south of Kodari, Nepal.[44][45] The aftershock caused fresh avalanches on Mount Everest and was felt in many places in northern India including Kolkata, Siliguri, Jalpaiguri and Assam.[46] The aftershock caused a landslide on the Koshi Highway which blocked the section of the road between Bhedetar and Mulghat.[47]

A model of Geo Gateway, based on a United States Geological Survey mechanism of a near-horizontal fault as well as location of aftershocks showed that the fault was an 11° dip striking at 295°, 50 km (31 mi) wide, 150 km (93 mi) long, and had a dip slip of 3 m (9.8 ft).[48] The USGS says the aftershock registered at a shallow depth of 10 km (6.2 mi).[46]

Assuming that 25 April earthquake was the largest event in this seismic episode, Nepal could expect more than 30 aftershocks greater than magnitude 5 over the following month.[49] As of 24 June 2015, 329 aftershocks had occurred with different epicenters and magnitudes equal to or above 4 M_w (out of which 47 aftershocks are equal to or above 5 M_w) and more than 20,000 aftershocks less than 4 M_w.[6]

6.3.5. 12 May 2015 Nepal Earthquake

A second major earthquake occurred on 12 May 2015 at 12:51 NST with a moment magnitude (M_w) of 7.3 M_w 18 km (11 mi) southeast of Kodari. The epicenter was near the Chinese border between the capital of Kathmandu and Mt. Everest. It struck at the depth of 18.5 km (11.5 miles). This earthquake occurred along the same fault as the original magnitude

7.8 earthquake of 25 April but further to the east. As such, it is considered to be an aftershock of the 25 April quake.[50] Tremors were also felt in northern parts of India including Bihar, Uttar Pradesh, West Bengal and other North-Indian States.[51][52][53]

At least 117 died in Nepal as a result of the aftershock and about 2,500 were injured. Seventeen others died in India and one in China. [24][54]

6.4.   Aftermath

Disastrous events in very poor and politically paralyzed nations such as Nepal often become a long drawn out chain of events, in that one disaster feeds into another for years or even decades upon end. The after effects from the earthquake have knock effects on a myriad seemingly unrelated aspects: human trafficking, labor cost and availability, rental and property cost burdens, urbanization, private and public debt burdens, mental health, politics, tourism, as well as disease and healthcare system damages, disasters that come with the monsoon season. The first monsoon related effects: a landslip on June 11th has claimed 53 lives[55] meanwhile a glacial lake had burst in particularly hard hit Solukhumbhu district;[56] whether or not the quake had contributed such events is often unknown and unresearched, but certainly possible.

Casualties by Country

Country	Deaths	Injuries	Ref.
Nepal	> 8,857	> 22,304	[7][57]
India	130	560	[58]
China	27	383	[59]
Bangladesh	4	200	[60]
Total	**9,018**	**23,447**	

Nepal

The earthquake killed more than 8,600 in Nepal[7][83] and injured more than twice as many. The rural death toll may have been lower than it would have been as the villagers were outdoors, working when the quake hit.[84] As of 15 May, 6,271 people, including 1,700 from the 12 May aftershock, were still receiving treatment for their injuries.[54] More than 450,000 people were displaced.[57]

The Himalayan Times reported that as many as 20,000 foreign nationals may have been visiting Nepal at the time of the earthquake, although reports of foreign deaths were relatively low.[85]

India

A total of 78 deaths were reported in India – 58 in Bihar, 16 in Uttar Pradesh, 3 in West Bengal and 1 in Rajasthan.[58]

China

25 dead and 4 missing, all from the Tibet Autonomous Region.[59]

Bangladesh

4 dead [60]

Foreign Casualties in Nepal

Country	Deaths	Ref.
India	40	[61]
France	10	[62]
Spain	7	[63]
United States	7	[64][65][66]
Germany	5	[67][68]
China	4	[69]
Italy	4	[70]
Canada	2	[71]
Russia	2	[72]
Australia India	1	[73][74]
Estonia	1	[75]
Hong Kong United Kingdom	1	[76][77]
Israel	1	[78]
Japan	1	[79]
Malaysia	1	[80]
New Zealand	1	[81]
United Kingdom	1	[82]
Total	**89**	

6.5.   Avalanches on Mount Everest

2015 Mount Everest Avalanches

This earthquake caused avalanches on Mount Everest. At least 19[86] died, including Google executive Dan Fredinburg,[87] with at least 120[86] others injured or missing.

6.6.   Landslides in the Langtang Valley

In the Langtang valley located in Langtang National Park 329 people were reported missing after an avalanche hit the village of Ghodatabela[88][89] and the village of Langtang. The avalanche was estimated to have been two to three kilometres wide. Ghodatabela was an area popular on the Langtang trekking route.[90] The village of Langtang has been destroyed by the avalanche. Smaller settlements on the outskirts of Langtang were buried during the earthquake, such as Chyamki, Thangsyap, and Mundu. Twelve locals and two foreigners were believed to have survived. Smaller landslides occurred in the Trishuli River Valley with reports of significant damage at Mailung, Simle, and Archale.[16][91][92] On 4 May it was announced that 52 bodies had been found in the Langtang area, of which seven were of foreigners.[93]

6.7.   Damages

Thousands of houses were destroyed across many districts of the country, with entire villages flattened, especially those near the epicenter.[15][17][18] The Tribhuvan International Airport serving Kathmandu was closed immediately after the quake, but was re-opened later in the day for relief operations and, later, for some commercial flights.[94] It subsequently shut down operations sporadically due to aftershocks,[95] and on 3 May was closed temporarily to the largest planes for fear of runway damage.[96] Many workers were not at their posts, either from becoming earthquake casualties or because they were dealing with its after effects.[97] Flights resumed from Pokhara, to the west of the epicenter, on 27 April.[98]

Image: Durbar Square

Durbar Square

Before Earthquake   After Earthquake

Building Damage as a Result of the Earthquake

Several of the churches in the Kathmandu valley were destroyed. As Saturday is the principal day of Christian worship in Nepal, 500 people are reported to have died in the collapses.[99][100]

Several pagodas on Kathmandu Durbar Square, a UNESCO World Heritage Site, collapsed, [26] as did the Dharahara tower, built in 1832; the collapse of the latter structure killed at least 180 people,[101][102][103][104] Manakamana Temple in Gorkha was also destroyed. The northern side of Janaki Mandir in Janakpur was reported to have been damaged.[105] Several temples, including Kasthamandap, Panchtale temple, the top levels of the nine-story Basantapur Durbar, the Dasa Avtar temple and two dewals located behind the Shiva Parvati temple were demolished by the quake. Some other monuments, including the Kumari Temple and the Taleju Bhawani Temple partially collapsed.[106][107]

The top of the Jaya Bageshwari Temple in Gaushala and some parts of thePashupatinath Temple, Swyambhunath, Boudhanath Stupa, Ratna Mandir, inside Rani Pokhari, and Durbar High School have been destroyed.[108]

In Patan, the Char Narayan Mandir, the statue of Yog Narendra Malla, a pati inside Patan Durbar Square, the Taleju Temple, the Hari Shankar, Uma Maheshwar Temple and the Machhindranath Temple in Bungamati were destroyed. In Tripureshwar, the Kal Mochan Ghat, a temple inspired by Mughal architecture, was completely destroyed and the nearby Tripura Sundari also suffered significant damage. In Bhaktapur, several monuments, including the Fasi Deva temple, the Chardham temple and the 17th century Vatsala Durga Temple, were fully or partially destroyed.[108]

Outside the Valley, the Manakamana Temple in Gorkha, the Gorkha Durbar, the Palanchok Bhagwati, in Kabhrepalanchok District, the Rani Mahal in Palpa District, the Churiyamai in Makwanpur District, the Dolakha Bhimsensthan in Dolakha District, and the Nuwakot Durbar were partially destroyed. Historian Prushottam Lochan Shrestha stated, "We have lost most of the monuments that had been designated as World Heritage Sites in Kathmandu, Bhaktapur and Lalitpur District, Nepal. They cannot be restored to their original states."[108] the north eastern parts of India also received major damage. Heavy shocks were felt including the states Uttrakhand, Uttar Pradesh, West Bengal and many other states. Huge damage was caused to the property and the lives of the people.

6.8. Economic Loss

Road Damaged in Nepal

Concern was expressed that harvests could be reduced or lost this season as people affected by the earthquake would have only a short time to plant crops before the onset of the Monsoon rains.[109]

Nepal, with a total Gross Domestic Product of USD$19.921 billion (according to a 2012 estimate),[110] is one of Asia's poorest countries, and has little ability to fund a major reconstruction effort on its own.[111] Even before the quake, the Asian Development Bank estimated that it would need to spend about four times more than it currently does annually on infrastructure through to 2020 to attract investment.[111] The U.S. Geological Survey initially estimated economic losses from the temblor at 9 percent to 50 percent of gross domestic product, with a best guess of 35 percent. "It's too hard for now to tell the extent of the damage and the effect on Nepal's GDP", according to Hun Kim, an Asian Development Bank (ADB) official. The ADB said on the 28th that it would provide a USD$3 million grant to Nepal for immediate relief efforts, and up to USD$200 million for the first phase of rehabilitation.[111]

Damaged House in Chaurikharka

Rajiv Biswas, an economist at a Colorado-based consultancy, said that rebuilding the economy will need international effort over the next few years as it could "easily exceed" USD$5 billion, or about 20 percent of Nepal's gross domestic product.[111][112][not in citation given] and #DontComeBackIndianMedia.[116]

6.9. Rescue and Relief

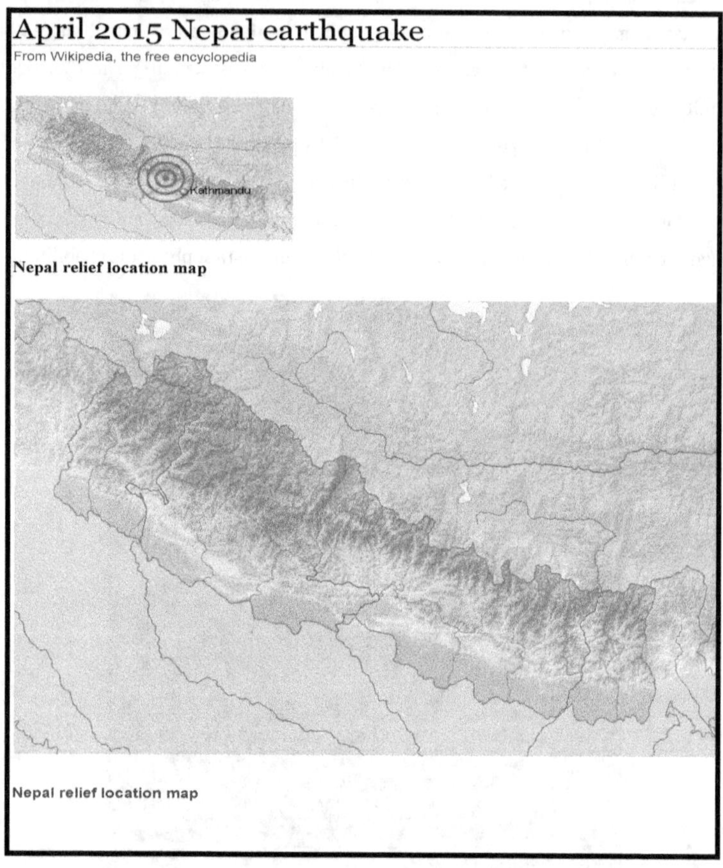

Map: Nepal Relief Locations

About 90 percent of soldiers from the Nepalese Army were sent to the stricken areas in the aftermath of the earthquake under Operation Sankat Mochan, with volunteers mobilized from other parts of the country.[117] Rainfall and aftershocks were factors complicating the rescue efforts, with potential secondary effects like additional landslides and further building collapses being concerns. Impassable roads and damaged communications infrastructure posed substantial challenges to rescue efforts.[118] Survivors were found up to a week after the earthquake.[119][120][121]

As of 1 May 2015, international aid agencies like Médecins Sans Frontières (Doctors Without Borders) and the Red Cross were able to start medically evacuating the critically wounded by helicopter from outlying areas, initially cut-off from the capital city, Kathmandu,[17] and treating others in mobile and makeshift facilities.[122][123] There was concern about epidemics due to the shortage of clean water, the makeshift nature of living conditions and the lack of toilets.[124]

Emergency workers were able to identify four men who had been trapped in rubble, and rescue them, using advanced heartbeat detection. The four men were trapped in up to ten feet of rubble in the village of Chautara, north of Kathmandu. An international team of rescuers from several countries using FINDER devices found two sets of men under two different collapsed buildings.[125]

Volunteers used crisis mapping to help plan emergency aid work.[126] Public volunteers from around the world added details into online maps.[127][128][129] Information was mapped from data input from social media, satellite pictures[130] anddrones[126] of passable roads, collapsed houses, stranded, shelterless and starving people, who needed help, and from messages and contact details of people willing to help.[131] On-site volunteers verified these mapping details wherever they could to reduce errors. First responders, from Nepali citizens to the Red Cross, the Nepal army and the United Nations used this data. The Nepal earthquake crisis mapping utilized experience gained and lessons learned about planning emergency aid work from earthquakes in Haiti and Indonesia.[132]

Reports are also coming in of sub-standard relief materials and inedible food being sent to Nepal by many of the foreign aid agencies.[133][134]

A United States Marine helicopter crashed on 12 May while involved in delivering relief supplies. The crash occurred at Charikot, roughly 45 miles (72 kilometers) east of Kathmandu. Two Nepali soldiers and 6 American soldiers died in the crash.[135]

6.10. Repair and Reconstruction

Monuments

UNESCO and the Ministry of Culture began strengthening damaged monuments in danger of collapsing before the monsoon season. Subsequent restoration of collapsed structures, including historic houses is planned. Architectural drawings exist that provide plans for reconstruction. According to UNESCO, more than 30 monuments in the Kathmandu Valley collapsed in the quakes, and another 120 incurred partial damage.[136] Repair estimates are

$160 million to restore 1,000 damaged and destroyed monasteries, temples, historic houses, and shrines across the country. The destruction is concentrated in the Kathmandu Valley. UNESCO designated seven groups of multi-ethnic monuments clustered in the valley as a single World Heritage Site, including Swayambhu, the Durbar squares of Kathmandu, Patan, and Bhaktapur, and the Hindu temples of Pashupati and Changu Narayan. Damaged in the quakes were the structures in the three Durbar squares, the temple of Changu Narayan, and the 1655 temple in Sankhu. Drones fly above a cultural heritage sites to provide 3D images of the damage to use for planning repairs. [137]

6.11. International Aid

UNICEF appealed for donations, as close to 1.7 million children had been driven out into the open, and were in desperate need of drinking water, psychological counsel, temporary shelters, sanitation and protection from disease outbreak. It distributed water, tents, hygiene kits, water purification tablets and buckets.[138] Numerous other organizations provided similar support.[139]

India was the first to respond within hours, being Nepal's immediate neighbour, [140] with Operation Maitri which provided rescue and relief by its armed forces. It also evacuated its own and other countries' stranded nationals. The United Kingdomhas been the largest bilateral aid donor to Nepal following the earthquake.[141][142] The United States, China and other nations have provided helicopters as requested by the Nepali government.[143][144]

On 26 April 2015, international aid agencies and governments mobilized rescue workers and aid for the earthquake. They faced challenges in both getting assistance to Nepal and ferrying people to remote areas as the country had few helicopters.[145][146] Relief efforts were also hampered by Nepalese government insistence on routing aid through the Prime Minister's Disaster Relief Fund and its National Emergency Operation Center. After concerns were raised, it was clarified that "Non-profits" or NGOs already in the country could continue receiving aid directly and bypass the official fund.[145][147]Aid mismatch and supply of "leftovers" by donors,[148] aid diversion in Nepal,[149] mistrust over control of the distribution of funds and supplies,[150][151][152] congestion and customs delays at Kathmandu's airport and border check posts were also reported.[153][154] On 3 May 2015, restrictions were placed on heavy aircraft flying in aid supplies after new cracks were noticed on the runway at the Tribhuvan airport (KTM), Nepal's only wide-body jet airport.[155][156][157]

6.12. 2004 Indian Ocean Earthquake and Tsunami

https://en.wikipedia.org/wiki/2004_Indian_Ocean_earthquake_and_tsunami

Facts

Date	26 December 2004[158]
Origin time	00:58 UTC
Magnitude	9.1 M_w[158]
Depth	30 km (19 mi)[158]
Epicenter	3.316°N 95.854°ECoordinates: 3.316°N 95.854°E[1]
Type	Megathrust
Areas affected	Indonesia Sri Lanka India Thailand Maldives Somalia
Tsunami	Yes
Casualties	230,000–280,000 dead and more missing[159][160][4161][162]

The **2004 Indian Ocean earthquake/tsunami** occurred at 00:58:53 UTC on 26 December with an epicenter off the west coast of Sumatra, Indonesia. The event is known by the scientific community as the Sumatra–Andaman earthquake.[163][164] The resulting tsunami was given various names, including the 2004 Indian Ocean tsunami, South Asian tsunami, Indonesian tsunami, the Christmas tsunami and the **Boxing Day tsunami**.

The undersea mega thrust earthquake was caused when the Indian Plate was sub ducted by the Burma Plate and triggered a series of devastating tsunamis along the coasts of most landmasses bordering the Indian Ocean, killing 230,000 people in 14 countries, and inundating coastal communities with waves up to 30 meters (100 ft) high. [165] It was one of the deadliest natural disasters in recorded history. Indonesia was the hardest-hit country, followed by Sri Lanka, India, and Thailand.

With a magnitude of M_w 9.1–9.3, it is the third-largest earthquake ever recorded on a seismograph. The earthquake had the longest duration of faulting ever observed, between 8.3 and 10 minutes. [166] It caused the entire planet to vibrate as much as 1 centimeter (0.4 inches) [167] and triggered other earthquakes as far away as Alaska.[168] Its epicenter was between Simeulue and mainland Indonesia.[169] The plight of the affected people and countries prompted a worldwide humanitarian response. In all, the worldwide community donated more than US\$14 billion (2004) in humanitarian aid. [170]

Important Facts Related to 2004 Indian Ocean Tsunami are Presented below: [171]

1. The December 26, 2004 Indian Ocean tsunami was caused by an earthquake that is thought to have had the energy of 23,000 Hiroshima-type atomic bombs.

2. The epicenter of the 9.0 magnitude quake was located in the Indian Ocean near the west coast of Sumatra.

3. The violent movement of the Earth's tectonic plates displaced an enormous amount of water, sending powerful shock waves in every direction.

4. The tectonic plates in this area had been pushing against each other and building pressure for thousands of years. They continue to do so and will likely cause underwater earthquakes and tsunamis in the future.

5. The rupture was more than 600 miles long, displacing the seafloor by 10 yards horizontally and several yards vertically. As a result, trillions of tons of rock moved, causing the largest magnitude earthquake in 40 years.

6. Within hours of the earthquake, killer waves radiating from the epicenter slammed into the coastlines of 11 countries, damaging countries from east Africa to Thailand.

7. A tsunami is a series of waves - the first wave may not be the most dangerous. A tsunami "wave train" may come in surges five minutes to an hour apart. The cycle is marked by the repeated retreat and advance of the ocean.

8. Despite a lag of up to several hours between the earthquake and the impact of the tsunami, nearly all of the victims were taken completely by surprise because there were no tsunami warning systems in place.

9. The Indian Ocean tsunami traveled as far as 3,000 miles to Africa and still arrived with sufficient force to kill people and destroy property.

10. Many people in Indonesia reported that they saw animals fleeing for high ground minutes before the tsunami arrived – very few animal bodies were found afterward.

11. The tsunami resulted in at least 227,898 fatalities.

The Facts as Stated above Amply Demonstrate, how Deadly the Tsunami was.

References

1. "M7.8 – 34 km ESE of Lamjung, Nepal". United States Geological Survey. 25 April 2015. Retrieved 12 May 2015.

2. China Earthquake Networks Center. 25 April 2015. Retrieved 28 April 2015.

3. Insurance in Asia: Narrow-minded, economist.com.

4. "M6.6 - 49km E of Lamjung, Nepal".

5. "M6.7 - 17km S of Kodari, Nepal".usgs.gov.

6. Aftershocks of Gorkha Earthquake". National Seismological Centre, Nepal.

7. "Incident Report of Earthquake 2015". Nepal Disaster Risk Reduction Portal.drrportal. gov.np. Retrieved28 May 2015.

8. "Nepal earthquake death toll rises to 8,413". The Times of India. 7 May 2015. Retrieved 9 May 2015.

9. Chidanand Rajghatta (26 April 2015). "Is this the 'Big Himalayan Quake' we feared?". The Times of India. Retrieved 26 April 2015.

10. "What 1934 Told Nepal to Expect About the Next Big Quake".

11. "Timeline: Nepal 2015 to 1934, the worst quake disasters in the last 80 years". Retrieved 28 April 2015.

12. "Nepal earthquake: Eerie reminder of 1934 tragedy".

13. nytimes.com 2015-04-27 katmandu-nepal-fear-loss-and-devastation, nytimes.com; accessed 28 April 2015.

14. "Trapped at 20,000 feet: Hundreds of Everest climbers await choppers as supplies run low". Fox News. The Associated Press. 26 April 2015. Retrieved 26 April 2015.

15. Shrestha, Sahina (1 May 2015). "Langtang is gone". Nepali Times. Retrieved 2 May 2015.

16. "Up to 250 missing after avalanche hits Nepal trekking route". "Mail Online". Retrieved 28 April 2015.

17. McCarthy, Julie (30 April 2015). "He Carried His Mom On His Back For 5 Hours En Route To Medical Care". Montana Public Radio. Retrieved 2 May 2015.

18. [b] Kaini, Sudip (29 April 2015). "Great Earthquake wipes out Barpak". The Kathmandu Post. Retrieved 2 May2015.

19. "Experts had warned for decades that Nepal was vulnerable to a killer quake". Washington Post. Retrieved29 April 2015.

20. Colin Stark (26 April 2015). "Nepal earthquake: 'A tragedy waiting to happen'- CNN. com". CNN. Retrieved 29 April2015.

21. "Earth Quake-hit Nepal at high risk of landslides in coming weeks". Free Press Journal.

22. "Earthquake Again in Nepal 12 May 2015". The Natural Disasters.

23. "Nepal earthquake, magnitude 7.3, strikes near Everest". BBC News. 12 May 2015.

24. Manesh Shrestha (14 May 2015). "Death toll from this week's Nepal earthquake rises above 125". CNN. Retrieved 14 May 2015.

25. Howard, Brian Clark (25 April 2015). "Nepal Earthquake Strikes One of Earth's Most Quake-Prone Areas". National Geographic. Retrieved 27 April 2015.

26. "Powerful earthquake hits Nepal". Al Jazeera. Retrieved 25 April 2015.

27. "1 Day, Magnitude 2.5+ Worldwide". Retrieved 25 April2015.

28. Joel Achenbach (27 April 2015). "Kathmandu shifts three metres in 30 seconds". Sydney Morning Herald (Washington Post). Retrieved 28 April 2015.

29. Internet Desk. "Was this the big earthquake that was predicted in the Himalayas?". The Hindu. Retrieved28 April 2015.

30. "Experts had warned for decades that Nepal was vulnerable to a killer quake". Washington Post. Retrieved27 April 2015.

31. Amos, Jonathan (25 April 2015). "Why Nepal is so vulnerable to quakes". BBC. Retrieved 27 April 2015.

32. "Why Earthquakes Are Devastating Nepal".video.nationalgeographic.com. Retrieved 13 May 2015.

33. "GEOLOGY OF NEPAL HIMALAYA". n.d. Retrieved27 April 2015.

34. "General Geology". Government of Nepal. 2014. Retrieved 27 April 2015.

35. Josh Fischman. "How The Deadly Nepal Earthquake Happened [GRAPHIC]". scientificamerican.com. Retrieved 28 April 2015.

36. Mughier J.L., Huyghe P., Gajurel A.P., Upreti B.N. & Jouanne F.; Huyghe; Gajurel; Upreti; Jouanne (2011)."Seismites in the Kathmandu basin and seismic hazard in central Himalaya" (PDF). Tectonophysics 509 (1–2): 33–49. Bibcode:2011Tectp.509...33M.doi:10.1016/j.tecto.2011.05.012.Retrieved 28april 2015.

37. L. Bollinger, S. N. Sapkota, P. Tapponnier, Y. Klinger, M. Rizza, J. Van der Woerd, D. R. Tiwari, R. Pandey, A. Bitri, S. Bes de Berc (2014). "Return period of great Himalayan earthquakes in Eastern Nepal: evidence from the Patu and Bardibas strands of the Main Frontal Thrust". Journal of Geophysical Research. Bibcode: 2014JGRB..119.7123B. doi: 10.1002/2014JB010970.

38. "Nepal quake 'followed historic pattern'". BBC. 27 April 2015.

39. "Andhra Pradesh earthquake update: Aftershocks of Nepal Earthquake felt in Visakhapatnam, Godavari and Srikakulam districts". Retrieved 28 April 2015.

40. "Tremors felt in Delhi, Metro services briefly disrupted".

41. "In pics: 5,500 killed, more than 1/4th of Nepal's population affected after horrific earthquake".

42. "Massive 7.9 Earthquake jolts Bihar, North India and Nepal". news.biharprabha.com. 25 April 2015. Retrieved25 April 2015.

43. China News. 25 April 2015. Retrieved 28 April2015.

44. Kim Hjelmgaard (26 April 2015). "Nepal hit by major aftershock as search for quake survivors intensifies".USA Today. Retrieved 26 April 2015.

45. "Fresh Tremors in North India, Including Delhi, a Day After Nepal Earthquake". NDTV. 26 April 2015. Retrieved26 April 2015.

46. LIVE: Death toll rises to 2200 in Nepal earthquake, India resumes rescue ops". The Indian Express. 26 April 2015. Retrieved 26 April 2015.

47. "Koshi Highway obstructed". ekantipur.com. Retrieved27 April 2015.

48. "IRIS: Special Event: Nepal". iris.edu.

49. "Major earthquake hits Nepal". Nature News & Comment. Retrieved 28 April 2015.

50. "M7.3 - 18km SE of Kodari, Nepal". USGS Earthquake Hazards Program.

51. "7.3 Magnitude Earthquake hits North India including Bihar". news.biharprabha.com. 12 May 2015. Retrieved12 May 2015.

52. "Nepal earthquake, magnitude 7.3, strikes near Everest". BBC News. 12 May 2015. Retrieved 12 May2015.

53. "M7.3-18km SE of Kodari, Nepal". United States Geological Survey. 12 May 2015. Retrieved 12 May 2015.

54. Manesh Shrestha; Jethro Mullen; Laura Smith-Spark (15 May 2015). "Nepal's latest earthquake: Death toll climbs above 100". CNN. Retrieved 15 May 2015.

55. "Taplejung landslide death toll reaches 53".ekantipur.com. Retrieved 13 June 2015.

56. "Glacial lake burst triggers floods in Solu".ekantipur.com. Retrieved 13 June 2015.

57. "National Emergency Operation Centre". National Emergency Operation Centre (Nepal Govt.) on Twitter (in Nepali). 3 May 2015. Retrieved 3 May 2015.

58. "Quake toll in India now 78". Zee News. 29 April 2015. Retrieved 4 May 2015.

59. "25 dead, 383 injured in Tibet following Nepal earthquake". Xinhua. 28 April 2015. Retrieved 5 May2015.

60. "4 killed, 18 Bangladesh districts affected in earthquake, says govt". Bdnews24.com. Retrieved26 April 2015.

61. "40-indian-nationals-confirmed-dead-in-nepalearthquake-8-are-missing". Retrieved 4 May 2015.

62. Ministère des Affaires étrangères et du Développement international. "Népal-Tremblement de terre -Point de situation (11.05.15)". France Diplomatie::Ministère des Affaires étrangères et du Développement international.

63. "España concluye el rescate en Nepal al dar por muertos a los desaparecidos" (in Spanish). El País. 9 May 2015. Retrieved 12 May 2015.

64. "The fourth American confirmed to have died was Vinh B Truong, who had been enjoying a hiking holiday when the tragedy happened; death toll soars past 5,000". Daily Mail. 27 April 2015. Retrieved 27 April 2015.

65. "Families now convinced 2 Seattle teens died in Nepal quake". KOMO News. Associated Press. 14 May 2015. Retrieved 15 May 2015.

66. "http://www.pressherald.com/2015/06/09/family-of-mainer-dawn-habash-acknowledges-her-death-on-facebook/". Portland Press Herald. 9 June 2015. Retrieved 13 June 2015.

67. "Foreigner victims Detail of Earthquake" (PDF). Nepal Police. Retrieved 3 May 2015.

68. "Fünftes deutsches Erdbeben-Opfer entdeckt". n24.de. 13 May 2015. Retrieved 14 May 2015.

69. "Nepal earthquake: International aid effort increased". BBC News. 27 April 2015. Retrieved 27 April 2015.

70. "terremoto Nepal". Emergenza24. Retrieved 27 April2015.

71. "St. Albert couple confirmed dead following Nepal earthquake". CTV Edmonton. 2 May 2015. Retrieved2 May 2015.

72. "Russian Diplomats Missing in Nepal Found Dead". The Moscow Times. 8 May 2015. Retrieved 8 May 2015.

73. "Kashmiri-origin mountaineer Renu Fotedar dies in Everest avalanche". Hindustan Times. 28 April 2015. Retrieved 2 May 2015.

74. "Nepal earthquake: Melbourne woman Renu Fotedar killed". The Sydney Morning Herald. 27 April 2015. Retrieved 28 April 2015.

- "Teadaolevalt on Nepali maavärinas hukkunud üks Eesti kodanik" [One Estonian citizen is known to have died in the Nepal earthquake] (in Estonian). Postimees. 26 April 2015. Retrieved 28 April 2015.

- "Earthquake claims life of one Estonian". Postimees. Retrieved 28 April 2015.

75. "Hong Kong resident confirmed dead in Nepal earthquake as 15 more remain missing". SCMP. 28 April 2015. Retrieved 29 April 2015.

76. "British death in Nepal earthquake confirmed". BBC News. 29 April 2015. Retrieved 3 May 2015.

- "Rescue team: Body of missing Israeli found in Nepal" [Emergency relief group says 22-year-old Or Asraf's body was located and identified]. The Times Of Israel. 3 May 2015. Retrieved 3 May 2015.

77. "Nepal earthquake death toll expected to rise sharply - live updates". The Guardian. 27 April 2015. Retrieved28 April 2015.

78. Desiree Tresa Gasper (8 May 2015). "Body confirmed to be Malaysian in Nepal earthquake". The Star/Asia News Network. AsiaOne. Retrieved 8 May 2015.

79. Andrea Vance; Rosanna Price (28 April 2015). "One Kiwi dead after Nepal earthquake". Stuff.co.nz. Retrieved28 April 2015.

80. "Nepal earthquake: British gap year student confirmed dead". BBC News. 9 May 2015. Retrieved 10 May 2015.

81. Jason Burke and Ishwar Rauniyar in Kathmandu (1 May 2015). "Nepal earthquake death toll exceeds 6,000 with thousands unaccounted for | World news". The Guardian. Retrieved 3 May 2015.

82. Juuphttps://web.archive.org/web/20150629024928/http://drrportal.gov.np/inciden treport

83. Jump up The Latest on Nepal: In Ravaged Hamlets, Lives Were Spared, U.S. News and World Reports, 29 April 2015

84. Jump up "90 Britons missing as Nepal earthquake death toll rises". The Himalayan Times via RT. 27 April 2015. Retrieved 27 April 2015.

85. Jump up to:[a] [b] "Everest's avalanche, through the climbers' eyes".Washington Post. 2 May 2015. Retrieved 4 May 2015.

86. Los Angeles Times (26 April 2015). "Google employee killed in avalanche was 'adventure activist,' friends say".latimes.com. Retrieved 29 April 2015.

87. "89 foreigners missing after earthquake in Nepal". aswani. 23 May 2015.

88. Daigle, Katy (28 April 2015). "250 people feared missing after mudslide in Nepal village". Daily News. Daily News. Retrieved 1 May 2015.

89. Dr Dave (29 April 2015). "Landslides in Langtang during and after the Nepal earthquake". AGU Blogosphere. AGU Blogosphere. Retrieved 1 May 2015.

90. "British archeologist 'terrified' after narrowly escaping death in Nepalese earthquake". Retrieved 28 April 2015.

91. "Landslides in Langtang during and after the Nepal earthquake". Rappler.com. Agence France-Presse. 29 April 2015. Retrieved 1 May 2015.

92. SHASHANK BENGALI AND BHRIKUTI RAI. "Nepal quake: 52 bodies found in remote valley where trekkers missing". Los Angeles Times. Retrieved 4 May 2015.

93. "Kathmandu airport shut, flights to resume from Sunday". ABP Live. 25 April 2015. Retrieved 26 April2015.

94. "Quake agony revealed quietly on trip from Nepal airport". Washington Post. Retrieved 28 April 2015.

95. "Kathmandu shuts airport to big jets". Irish Examiner. Retrieved 4 May 2015.

96. "In earthquake's deadly aftermath, Nepalis grieve the loss of their sacred landmarks". Washington Post. Retrieved28 April 2015.

97. "Kathmandu shuts airport to big jets". Irish Examiner. Retrieved 4 May 2015.

98. "In earthquake's deadly aftermath, Nepalis grieve the loss of their sacred landmarks". Washington Post. Retrieved28 April 2015.

99. "Relief for tourists after Pokhara flights resume".Myanmar Times.

100 "Nepal Quake Collapses Churches on Worshippers During Service; Bodies Still Not Recovered". Christianpost.com. Retrieved 3 May 2015.

101 "Nepal Earthquake Collapses Churches during Weekly Worship Services". Christianity Today.com. Retrieved3 May 2015.

102 "Historic Tower Collapses In Nepal Earthquake". The Huffington Post. Retrieved 28 April 2015.

103 Deepak Nagpal (25 April 2015). "LIVE: Two major quakes rattle Nepal; historic Dharahara Tower collapses, deaths reported in India". Zee News. Retrieved 25 April 2015.

104 "Historic Dharahara tower collapses in Kathmandu after earthquake". 25 April 2015. Retrieved 28 April 2015.

105 "180 bodies retrieved from debris of Nepal's historic tower". Retrieved 28 April 2015.

106 "Nearly 700 killed after 7.9-magnitude earthquake strikes Nepal". RT. 25 April 2015. Retrieved 25 April 2015.

107 "Earthquake in Nepal 2015". The Natural Disasters.

108 Before and After the Earthquake in Nepal (photographs), nytimes.com, 25 April 2015; accessed 4 May 2015.

109 Jump up to:[a][b][c] "Historical monuments lost forever". The Nation. 26 April 2015. Retrieved 28 April 2015.

110 "In Nepal, senior UN official warns 'clock is ticking' for earthquake relief efforts". The United Nations News Centre. 4 May 2015. Retrieved 4 May 2015.

111 "Nepal Economy Devastated Following Earthquake".PrimePair. Retrieved 27 April 2015.

112 "Nepal's Slowing Economy Set for Freefall Without Global Help". Bloomberg Business. Retrieved 26 April2015.

113 "The Latest on Nepal Quake: Aid arriving as deaths top 4,000". The New Indian Express. Retrieved 4 May 2015.

114 Burke, Jason (5 May 2015). "Nepal quake survivors face threat from human traffickers supplying sex trade". The Guardian. Retrieved 7 May 2015.

115 "Migrant Workers and their Families". United Nation Nepal Information Platform. Retrieved 7 May 2015.

116 http://www.ekantipur.com/2015/07/31/national/women-have-little-access-to-relief-report/408699.html Ekantipur Women have little access to relief: Report

117 http://www.ekantipur.com/2015/07/30/national/malnutrition-stalks-quake-hit-kids/408642.html Ekantipur Malnutrition stalks quake-hit kids

118 Sriram, Jayant (5 May 2015). "Indian media jingoism was trigger for backlash in Nepal". The Hindu. Retrieved5 May 2015.

119 "Nepal Earthquake: On World Press Freedom Day, #GoHomeIndianMedia top Twitter trend". dna webdesk. 4 May 2015. Retrieved 4 May 2015.

120 Deadly earthquake: Death-toll reaches 4400. In:Kantipur (daily), 28 April 2015.

121 J (A collection of Field Reports) (2 May 2015). "The Latest on Nepal: Chance of finding more survivors slim". The Tampa Tribune. Retrieved 2 May 2015.[dead link]

122 "Nepal Earthquake: Man Explains How He Survived in Rubble for 82 Hours". ABC News (US). AP. 29 April 2015. Retrieved 4 May 2015.

123 "Four survivors pulled alive from earthquake rubble in Nepal". RTE News, Ireland. 3 May 2015. Retrieved 4 May2015.

124 Sedhai, Roshan. "Still finding survivors".ekantipur.com. Retrieved 2 May 2015.

125 "Nepal-1 May 2015". Medecins Sans Frontieres (Doctors Without Borders). Retrieved 1 May 2015.

126 30 April 2015. "Japanese Red Cross medical teams reach communities beyond Kathmandu". International Federation of Red Cross and Red Crescent Societies. Retrieved 1 May 2015.

127 Lee, Brianna (30 April 2015). "Nepal Earthquake: In Tent Cities, Water Shortages, Open Toilets Add To Fears Of Looming Health Disaster". International Business Times. Retrieved 2 May 2015.

128 Howard, Brian Clark; 07, National Geographic PUBLISHED May. "NASA Technology Finds Nepal Survivors by Their Heartbeats". National Geographic News. Retrieved 8 May 2015.

129 Parker, Laura (1 May 2015). "How 'Crisis Mapping' Is Shaping Disaster Relief in Nepal". The National Geographic. Retrieved 4 May 2015.

130 "The Humanitarian OpenStreetMap Team (HOT)".

131 Clark, Liat (28 April 2015). "How Nepal's Earthquake Was Mapped in 48 Hours". Wired (U.K.). Retrieved 4 May2015.

132 Shakya, Ayesha (1 May 2015). "Mapping the aftermath". The Nepali Times. Retrieved 4 May 2015.

133 "Earthquake in Nepal, India, Bangladesh, China". un-spider.org. Retrieved 28 May 2015.

134 "Showing Reports... (Nepal Earthquakes 2015)".Kathmandu Living Labs. Retrieved 4 May 2015.

135 Soden, Robert; Budhathoki, Nama; Palen, Lasia."Resilience-Building and the Crisis Informatics Agenda: Lessons Learned from 'Open Cities Kathmandu'"(PDF). Proceedings of the 11th International ISCRAM Conference - University Park, Pennsylvania (USA) - (May 2014). Retrieved 4 May 2015.

136 India's Help |url=https://www.telegraphindia.com/1150625/jsp/frontpage/story_27 732.jsp#.VhX40H0xFee

137 "NHRC to govt: Take action against WFP".ekantipur.com. Retrieved 28 May 2015.

138 "7,000 liters of relief oil disposed". ekantipur.com. Retrieved 28 May 2015.

139 Sugam Pokharel and Kimberly Hutcherson. "8 bodies recovered from U.S. helicopter crash site; pilot identified".CNN. Retrieved 16 May 2015.

140 Romey, Kristin;27, National Geographic PUBLISHED April. "Nepal's 8 Key Historic Sites: What's Rubble, What's Still Standing". National Geographic News. Retrieved2015-06-10.

141 Lawler, rew; 08, National Geographic PUBLISHED June."A Race to Fix Nepal's Ravaged Monuments Before Monsoons Hit". National Geographic News. Retrieved2015-06-10.

142 "29 April 2015 - Nepal Earthquake: New appeal for children amid growing need- UNICEF". UNICEF (United Nations Children's Fund). Retrieved 30 April 2015.

143 "Open Sans". earthquakenepalin2015.com. Retrieved17 May 2015.

144 Bhardwaj, Mayank, Dutta, Ratnajyoti (28 April 2015)."China? India? We're grateful for their help: Nepal's ambassador to India". Reuters. Retrieved 30 April 2015.

145 "Nepal earthquake: UK aid response". GOV.UK. 14 May 2015. Retrieved 23 May 2015.

146 "RAF Chinooks recalled from Nepal quake effort without flying a mission". The Guardian. 15 May 2015. Retrieved16 May 2015.

147 "Nepal earthquake: Racing against time, government pleads for rescue helicopters to reach remote mountain regions". The Independent. 23 May 2015. Retrieved23 May 2015.

148 "Chinese helicopters aid Nepal quake recovery".Ministry of National Defense of the People's Republic of China. Xinhua/Wang Shoubao. 6 May 2015. Retrieved23 May2015.

149 Mark Scott. "Nepal Earthquake Poses Challenge to International Aid Agencies". The New York Times. Retrieved 28 April 2015.

150 Sifferlin, Alexandra (28 April 2015). "Medics Race Against Time to Save Nepal's Quake Survivors". TIME. Retrieved 30 April 2015.

151 Anyadike, Obinna (5 May 2015). "Nepal quake fund move is PR fiasco". IRIN Nairobi. Retrieved 6 May 2015.

152 Kumar, Ruchir (6 May 2015). "No, thank you: Nepal asks India to not send old clothes as relief". Hindustan Times. Retrieved 6 May 2015.

153 Ghimire, Yubaraj; Khan, Hamza; Dutta, Amrita (5 May 2015). "Nepal aid getting 'hijacked': Organised groups are obstructing the work or 'hijacking' aid material to Nepal midway.". The Indian Express. Retrieved 6 May 2015.

154 Daniel, Frank Jack; Mahr, Krista (6 May 2015). "Nepal, aid agencies trade blame as confusion mars quake relief". The Sydney Morning Herald. Retrieved 6 May2015.

155 Bell, Thomas (3 May 2015). "Nepal earthquake: Many survivors receiving no help despite relief effort". The Independent. Retrieved 3 May 2015.

156 "Satisfied with coordination of foreign relief aid, says Army chief". eKantipur. 2 May 2015. Retrieved 2 May2015.

157 Ratcliffe, Rebecca (2 May 2015). "Nepal customs holding up earthquake relief efforts, says United Nations". The Guardian. Retrieved 2 May 2015.

158 "Magnitude 9.1 – OFF THE WEST COAST OF SUMATRA". U.S. Geological Survey. Retrieved 26 August2012.

159 If the death toll in Myanmar was 400–600 as claimed by dissident groups there, rather than just 61 or 90, more than 230,000 people would have perished in total from the tsunami.

160 Earthquakes with 50,000 or More Deaths[dead link]

161 "Indonesia quake toll jumps again". BBC News. 25 January 2005. Retrieved 24 December 2012.

162 Indian Ocean tsunami anniversary: Memorial events held 26 December 2014, BBC News

163 Lay, T., Kanamori, H., Ammon, C., Nettles, M., Ward, S., Aster, R., Beck, S., Bilek, S., Brudzinski, M., Butler, R., DeShon, H., Ekström, G., Satake, K., Sipkin, S., The Great Sumatra-Andaman Earthquake of 26 December 2004,Science, 308, 1127–1133, doi:10.1126/science.1112250, 2005

164 "Tsunamis and Earthquakes: Tsunami Generation from the 2004 Sumatra Earthquake- USGS Western Coastal and Marine Geology". Walrus.wr.usgs.gov. Retrieved 12 August 2010.

165 Paris,R.; Lavigne F., Wassimer P. & Sartohadi J.(2007). "Coastal sedimentation associated with the December 26, 2004 tsunami in Lhok Nga, west Banda Aceh (Sumatra,Indonesia)". Marine   Geology (Elsevier) 238 (1–4):   93–106.doi:10.1016/ j.margeo.2006.12.009.

166 "Analysis of the Sumatra-Andaman Earthquake Reveals Longest Fault Rupture Ever"

167 Walton, Marsha. "Scientists: Sumatra quake longest ever recorded." CNN. 20 May 2005

168 West, Michael; Sanches, John J.; McNutt, Stephen R. "Periodically Triggered Seismicity at Mount Wrangell, Alaska, After the Sumatra Earthquake." Science. Vol. 308, No. 5725, 1144–1146. 20 May 2005.

169 Nalbant, S., Steacy, S., Sieh, K., Natawidjaja, D., and McCloskey, J. "Seismology: Earthquake risk on the Sunda trench." Nature. Vol. 435, No. 7043, 756–757. 9 June 2005. Retrieved 16 May 2009. Archived 18 May 2009.

170 Jayasuriya, Sisira and Peter McCawley, "The Asian Tsunami: Aid and Reconstruction after a Disaster". Cheltenham UK and Northampton MA USA: Edward Elgar, 2010.

171 https://www.dosomething.org/facts/11-facts-about-2004-indian-ocean-tsunami

7. Images-Characteristic Features of Earthquakes

Earthquakes are the phenomena experienced during sudden movements of the Earth's crust. Under the Earth's crust lies the astheno-sphere, the upper part of the mantle composed of liquid rock. The plates of the Earth's crust essentially "float" on top of this layer, and can be forced to shift as the upwelling molten material below moves.

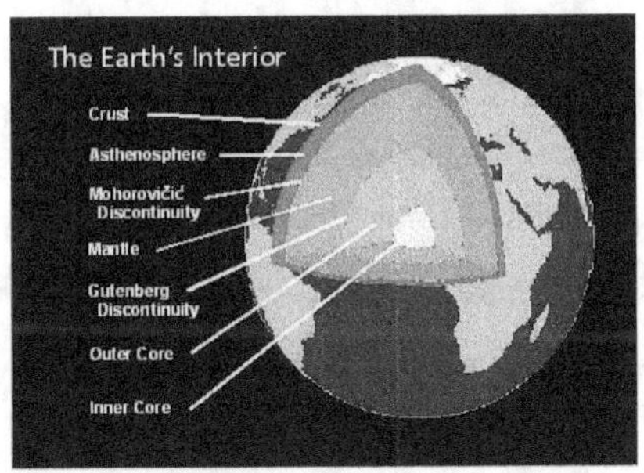

(Credit: Dylan Prentiss, UC-Santa Barbara) [1]

As the plates shift (and thus interact with each other), an enormous amount of energy is released in the form of waves. Although they can occur anywhere on the planet with little or no warning, the most extreme earthquakes occur near plate boundaries, as the plates converge (collide), diverge (move away from another), or shear (grind past one another). Moving rock and magma within volcanoes can also trigger earthquakes. In all of these cases, large sections of the crust can fracture and move to-and-fro to dissipate the released energy. This "shaking" is the sensation felt during an earthquake. The energy released is often described in terms of "magnitude," a logarithmic scale used to describe how energetic an earthquake was; a quake of magnitude 2 is hardly noticeable without special monitoring equipment, while quakes over magnitude 8 may actually cause the ground to visibly heave and roll. Since the scale is logarithmic, a magnitude 8 quake is not four times more energetic than a magnitude 2 quake, but one billion times more energetic!

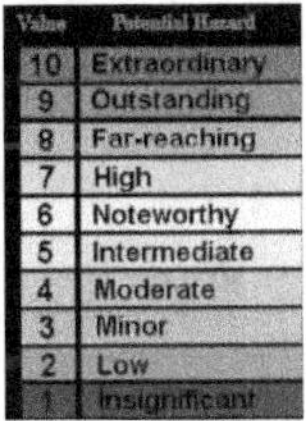

Value	Potential Hazard
10	Extraordinary
9	Outstanding
8	Far-reaching
7	High
6	Noteworthy
5	Intermediate
4	Moderate
3	Minor
2	Low
1	Insignificant

Image: Earthquake Intensity and Potential Hazard

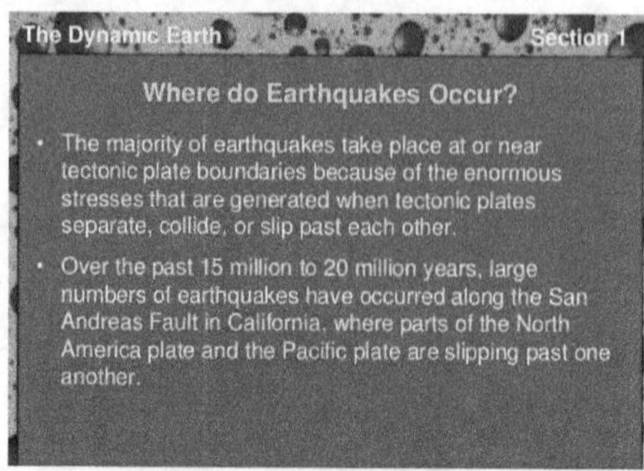

Earthquake prone Point and Place of Occurrence

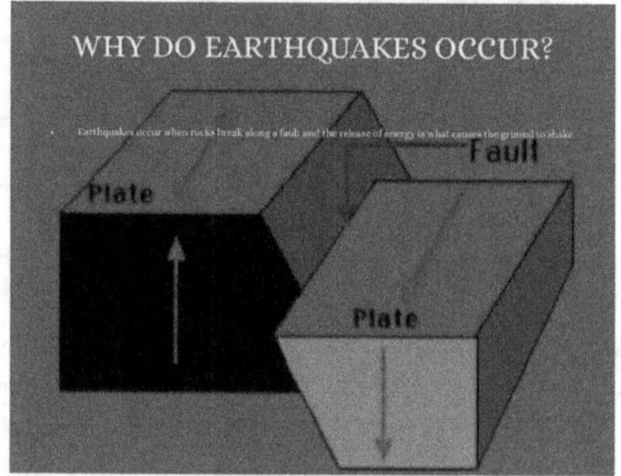

Earthquakes occur when rocks break along a fault and the release of energy is what causes the ground to shake [2]

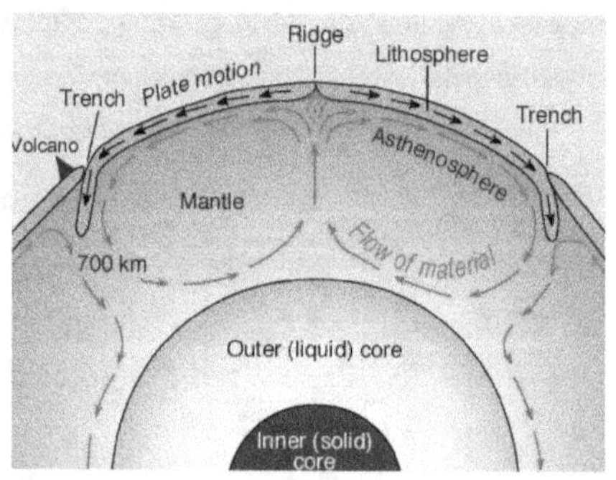

Image: Cut away earth's layers [3]

Cut away Earth's Layers

Earthquakes are the phenomena experienced during sudden movements of the earth's crust. Under the earth's crust lies the asthenosphere, the upper part of the mantle composed of liquid rock. The plates of the earth's crust essentially 'float' on top of this layer, and can be forced to shift as the upwelling molten material below moves. As the plates shift (and thus interact with each other), an enormous amount of energy is released in the form of waves. Although it can occur anywhere on the planet with little or no warning, the most extreme earthquakes occur near plate boundaries, as the plates converge (collide), diverge (move away from another), or shear (grind past one another).

Moving rock and magma within volcanoes can also trigger earthquakes. In all these cases large sections of the crust can fracture and move to and fro to dissipate the released energy. This 'shaking' is the sensation felt during an earthquake. The energy released is often described in terms of 'magnitude' (or Richter scale) a logarithmic scale used to describe how energetic an earthquake was. A quake of magnitude 2 is hardly noticeable without special monitoring equipment, while quakes over magnitude 8 may actually cause the ground to visibly heave and roll. Since the scale is logarithmic, a magnitude 8 quake is not for times more energetic than a magnitude 2 quake, but one billion times more energetic!

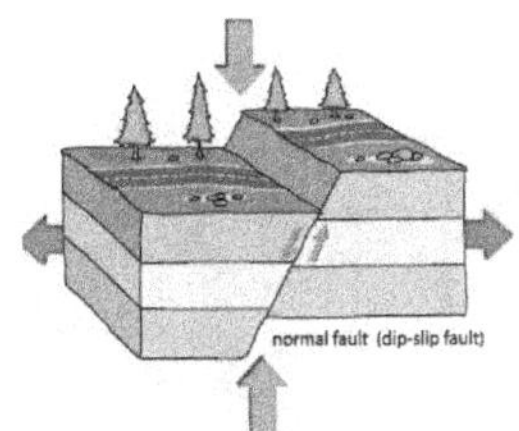

Image: A type of fault where the hanging wall slides downward relative to footwall caused by tension in the crust. Accommodate lengthening, extensional forces [4]

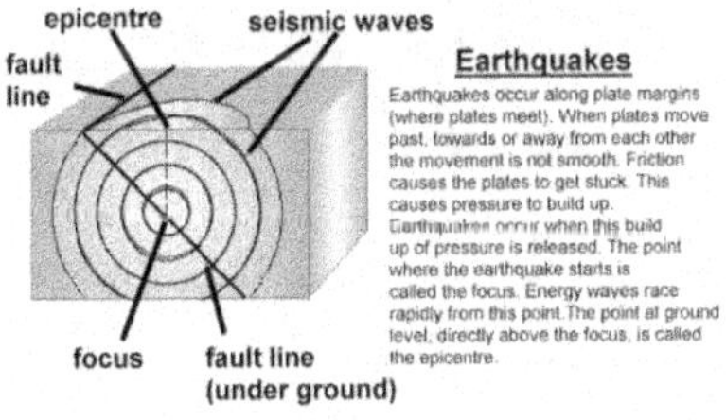

Earthquakes [5]

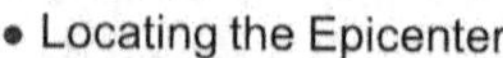
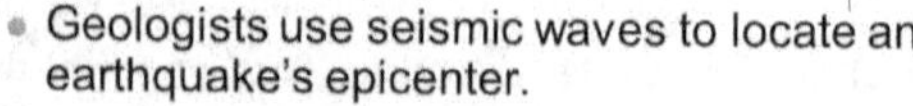

Earthquake and Seismic Waves

- Locating the Epicenter
  - Geologists use seismic waves to locate an earthquake's epicenter.
  - P waves arrive at a seismograph first, with S waves following close behind.
  - To tell how far the epicenter is from the seismograph, scientists measure the difference between the arrival times of the P waves and S waves.
  - The farther away an earthquake is, the greater the time between the arrival of the P waves and the S waves.

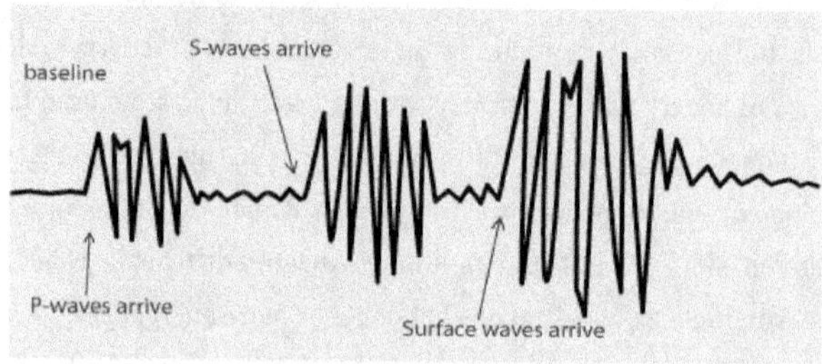

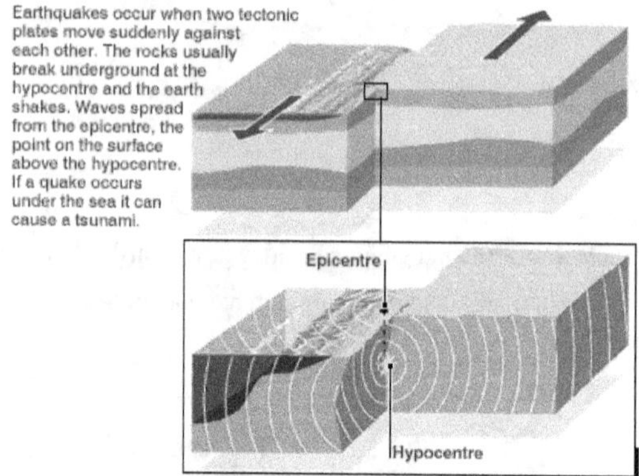

Image: Two tectonic plates moving suddenly against each other resulting in earthquake [6]

Other plates move slowly alongside each other: Faults are found at the edges of the plates where the crust is moving in different directions

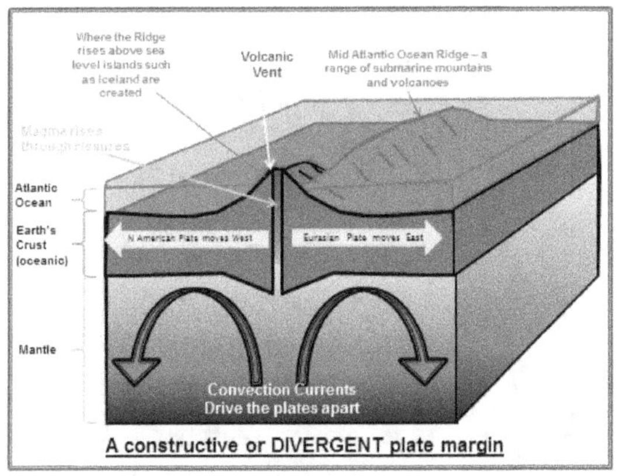

Divergent Plate Margin [7]

Strike-slip Faults [8]

Dominant along transform plate boundaries, *strike-slip faults* allow for sliding and displacement of adjacent landscapes. [8]

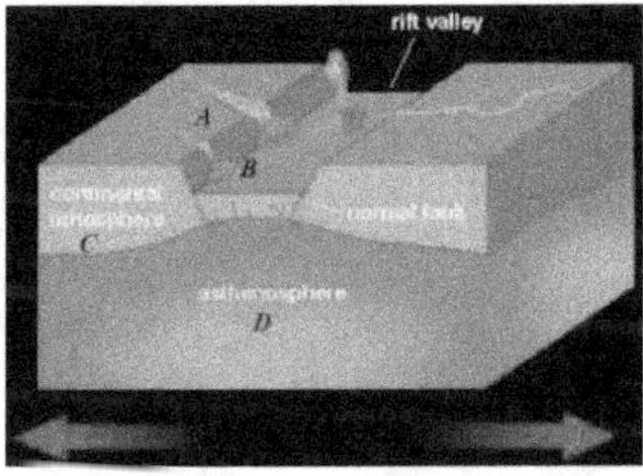

Rift-valley- Formation- leads-to- Earthquakes_ 5106 [9]

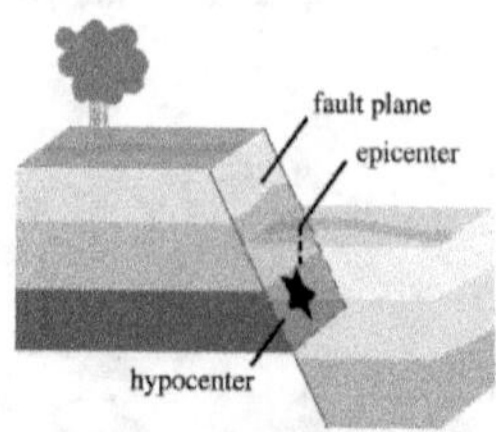

Image: fault plane, epicenter and hypocentre

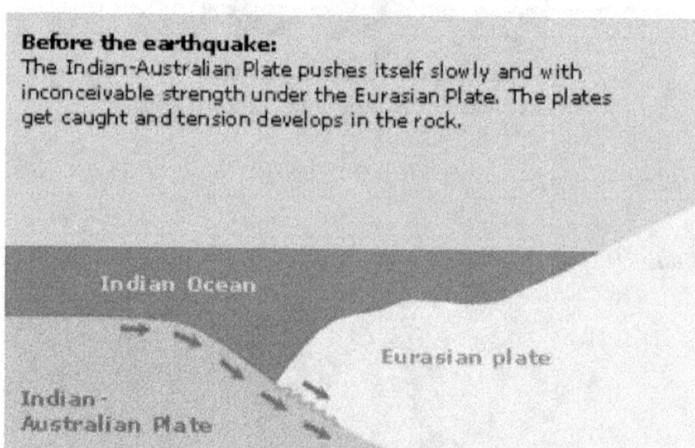

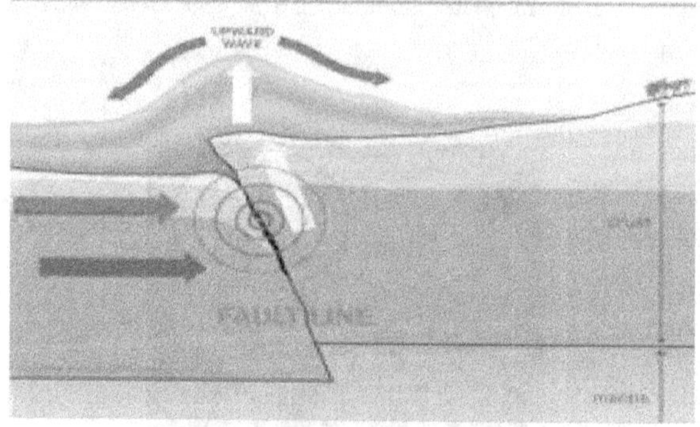

Indian Ocean Tsunami [10]

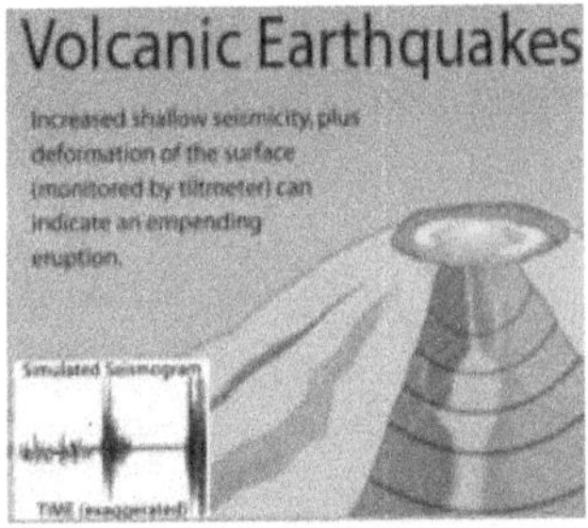

Volcanic Earthquakes [11]

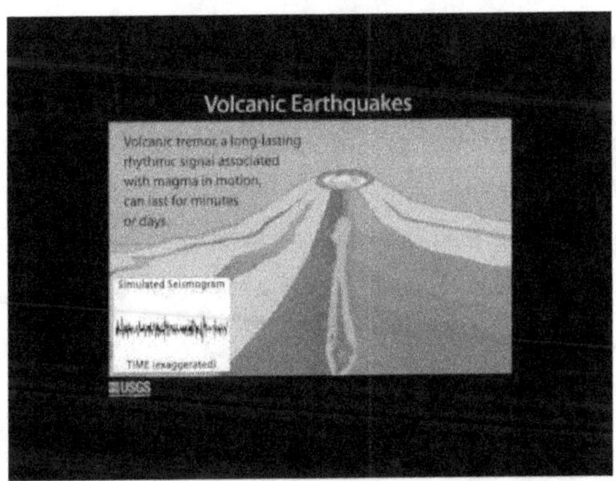

Volcanic Earthquakes [12]

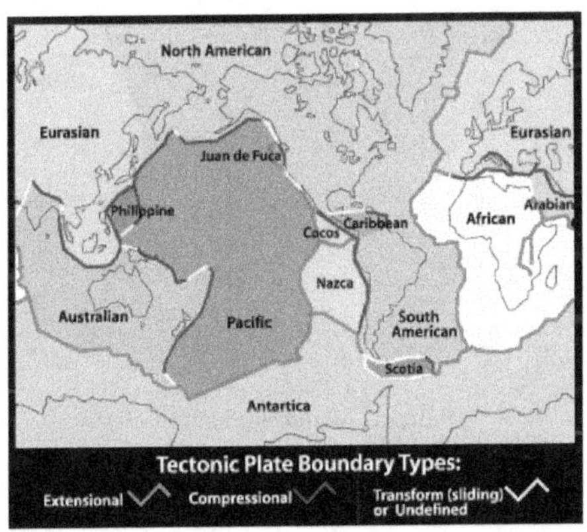

Plate Tectonics [13]

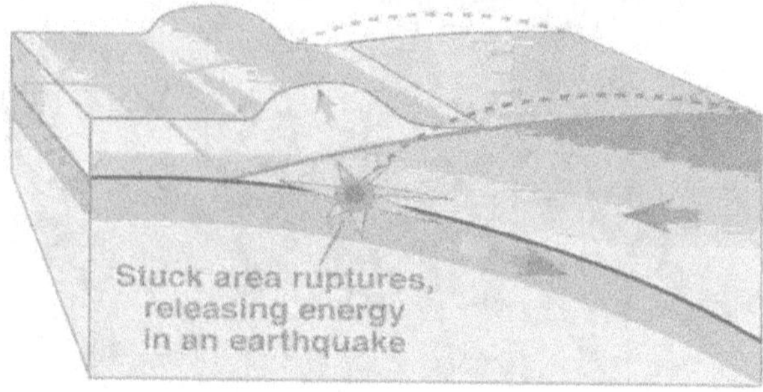

Diagram showing how Rupture along a Subduction Zone displaces the Seafloor and Overlying Sto form a Tsunami. [14]

Energy accumulates in the overriding plate until it exceeds the frictional forces between the two stuck plates. When this happens, the overriding plate snaps back into an unrestrained position. This sudden motion is the cause of the tsunami - because it gives an enormous shove to the overlying water. At the same time, inland areas of the overriding plate are suddenly lowered.

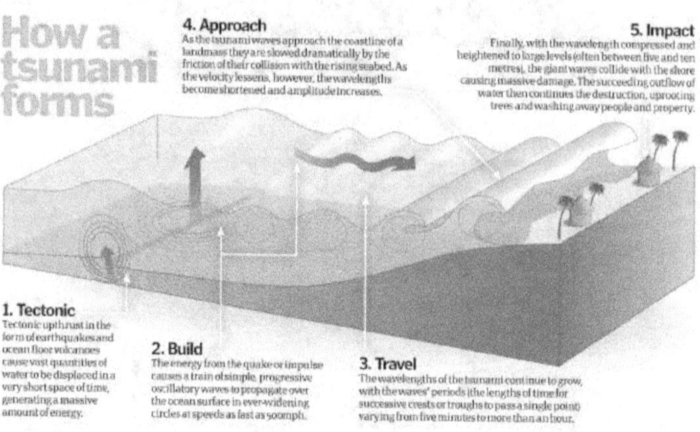

Tsunami Formation Large [15]

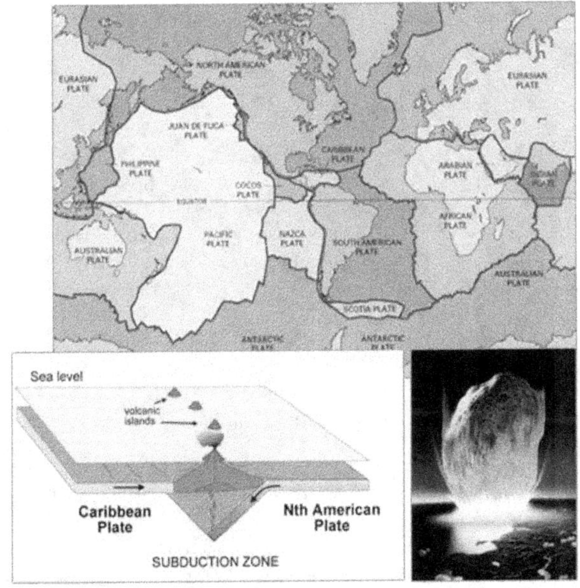

Tsunami Image Smart [16]

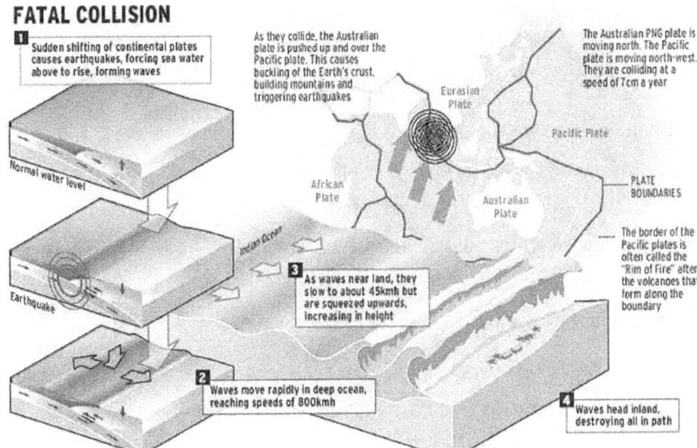

Tectonic Plates-Fatal Collision [17]

How does earthquake occurs?

+ Most earthquakes happen along the edge of the oceanic and continental plates. [Oceanic plates are under water, continental plates are above.] The earth's crust (the outer layer of the planet) is made up of a bunch of pieces, called plates. The plates get moved around by the liquid layers of magma underneath the Earth's crust. The plates are always bumping into each other, and pulling away from each other or past each other. Earthquakes usually happen when two plates are running into each other or sliding past each other. They can also happen along faults.

How does earthquake occur [18]

Earthquakes are usually caused when rock underground suddenly breaks along a fault. This sudden release of energy causes the seismic waves. It make the ground shake. When two blocks of rock or two plates are rubbing against each other, they stick a little. They don't just slide smoothly. The rocks are still pushing against each other, but not moving. After a while, the rocks break because of all the pressure that's built up.

When the rocks break, the earthquake occurs. During the earthquake and afterward, the plates or blocks of rock start moving, and they continue to move until they get stuck again.

Earthquake's Principle [19]

For understanding the earthquake; earth is generally considered s static and earthquakes are vibrations of the Earth caused by a rapid release of energy along a **fault** (a large fracture in the Earth's crust). Rocks along a fault line are bending and storing elastic energy. This strain is released through slippage which allows the rock to "snap back" as it elastically returns to its original shape in a process termed **elastic rebound**. This released energy radiates in all directions from its weakened source within the Earth, the **focus**, in the form of waves. [24]

Seismic waves are elastic energy and are characterized by surface waves or body waves and constantly inform geophysicists on the nature of Earth's interior.

Surface waves travel along the surface of the Earth. **Body waves** travel through Earth's interior and are divided into two types. **Primary waves** (P waves) are "push-pull," operating by compression and expansion, travel at high velocities and can travel through solids, liquids and gasses. **Secondary Waves** (S waves) are characterized by a "shaking" motion, travel at slower velocities and only through solids.

The **epicenter** is the area on the surface of the Earth which is directly above the focus and is located by using the difference in velocities of primary and secondary waves. A travel-time graph is used to calculate each of the stations' distance to the epicenter. A circle with a radius equal to the distance to the epicenter is drawn around each station. The point where all three circles intersect is the earthquake epicenter.

The graph on which seismic waves are recorded is called a seismogram. The amplitude of the recorded seismic wave is the vertical distance between the crest and trough of the waveform, therefore, the larger the earthquake, the greater the amplitude of the earthquake. The key to locating an earthquake's epicenter is the difference in arrival time, called lag time, of P- and S-waves.

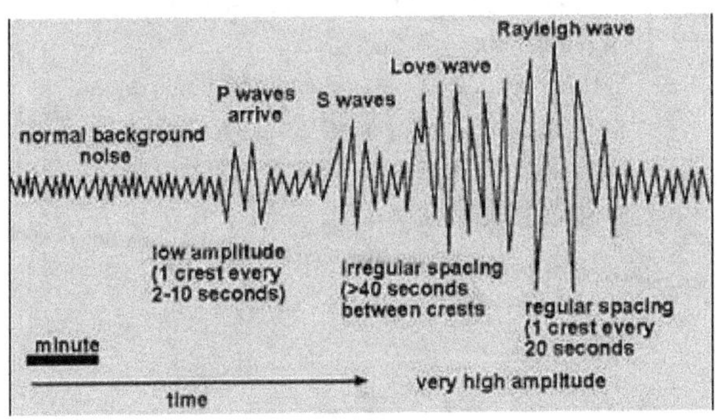

Schematic of a Seismogram [20]

The P waves and S waves are recorded on seismographs. Scientists use the difference in arrival times between the two sets of waves to determine the distance from where the rocks fractured at the beginning of the earthquake. Seismologists draw circles around the location of each reporting station. They use the time travel difference between the P waves and S waves to determine how far away the earthquake's focus is from each reporting station with a seismograph.

Seismologists gather data from at least three seismographs to plot the location of an earthquake. The point where at least three circles intersect on a map is the epicenter of the earthquake (Source: http://www.kids-earth-science.com/earthquake-epicenter.html)

It is important to people hearing the reports of a major earthquake to know approximately where the earthquake is located. Scientists know people can get a better understanding where an earthquake occurred if they give the information relative to towns and cities in an area. [26]

For instance, the epicenter of an earthquake might occur 20 miles north of Los Angeles, California. People hearing the news about the earthquake know it occurred in California. It probably occurred somewhere along or near the San Andreas Fault. Finally, they know the earthquake was near the metropolitan area of Los Angeles, California.

Locating an Earthquake

- ◆ Earthquake Distance
  - The epicenter is located using the difference in the arrival times between P and S wave recordings, which are related to distance.
- ◆ Earthquake Direction
  - Travel-time graphs from three or more seismographs can be used to find the exact location of an earthquake epicenter.
- ◆ Earthquake Zones
  - About 95 percent of the major earthquakes occur in a few narrow zones.

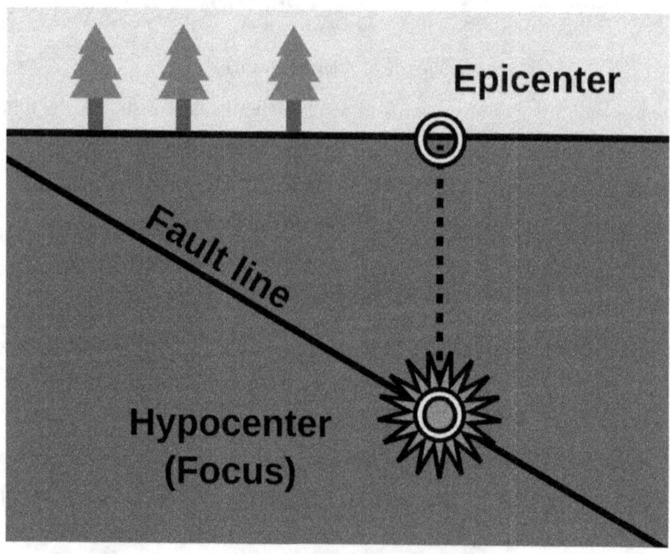

Schematic of Epicenter [21]

Locating the Epicenter

Geologists use seismic waves to locate an earthquake's epicenter.

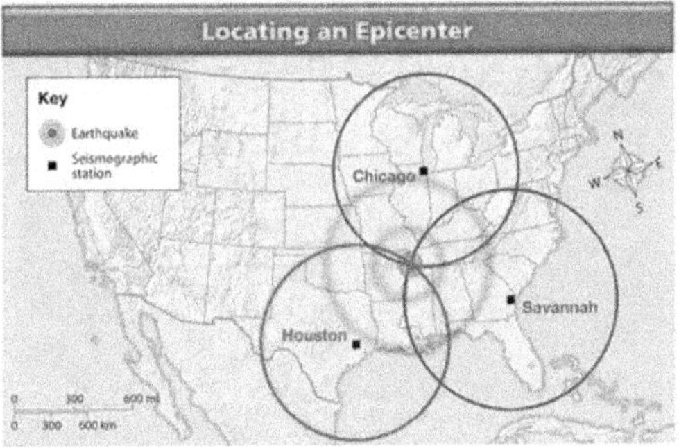

Locating the Epicenter [22]

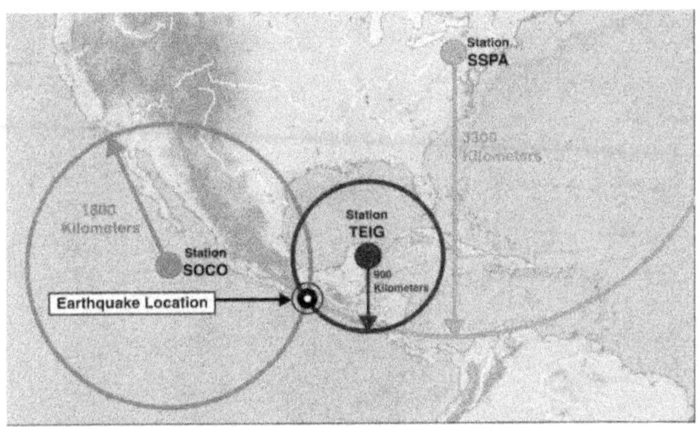

Earthquake Location from Three Stations [23]

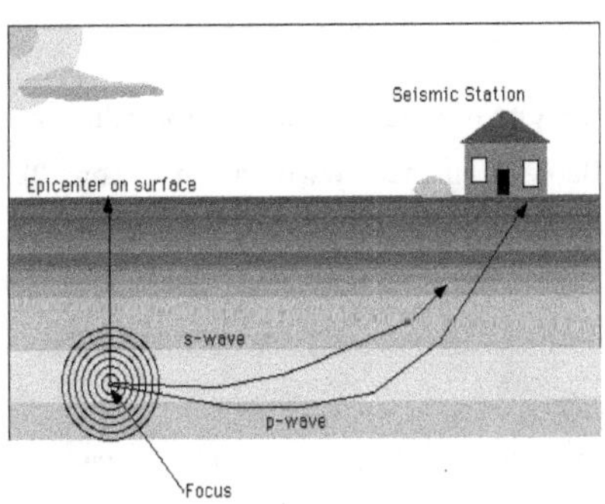

Epicenter from a Seismic Station [24]

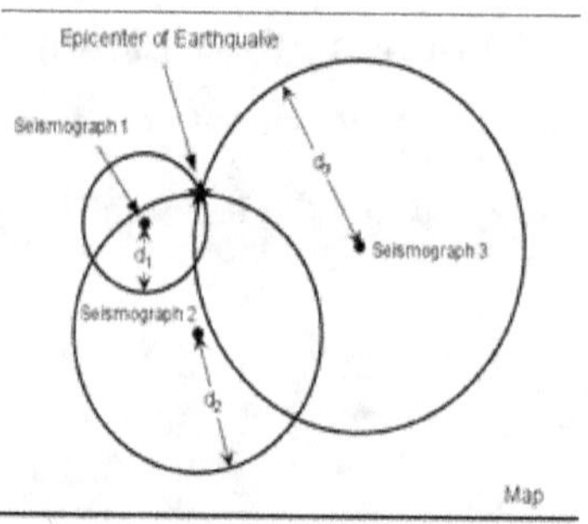

Map Showing Epicenter of an Earthquake [25]

Figure - the point where the three circles intersect is the epicenter of the earthquake. This technique is called 'triangulation [26]

References

1. http://earth.rice.edu/mtpe/geo/geosphere/hot/3earthquakes.html
2. Earthquakes by Hadley Hartmetz www.haikudeck.com1024 × 768Search by image Why do earthquakes occur
3. http://earth.rice.edu/Earth_Update/updating/geosphere/hot/cutaway_final.jpg
4. https://classconnection.s3.amazonaws.com/327/flashcards/538327/jpg/reverse_faul t1306091462200.jpg
5. http://www.geography.learnontheinternet.co.uk/images/nathaz/earthquake.gif
6. http://home.southernct.edu/~schmiedelt1/Image/earthquake.gif

7. http://www.coolgeography.co.uk/GCSE/AQA/Restless%20Earth/Tectonics/Constructive_Margin.png

8. *http://www.ei.lehigh.edu/learners/tectonics/faults/faults1.htm*

9. http://mail.colonial.net/~hkaiter/Aaa_web_images2012/arrift-valley-formation-leads-to-earthquakes_5106.jpg

10. https://lh5.googleusercontent.com/-aLAitHDEztI/TYbKQzBjOmI/AAAAAAAAAxU/YdXfWgauAR0/s1600/Tsunami.bmp

11. http://www.iris.edu/hq/inclass/uploads/Monitor%20volc%20seis_thumbnail.jpg

12. https://i.ytimg.com/vi/JypTLDLABzM/maxresdefault.jpg

13. http://mail.colonial.net/~hkaiter/platetectonics.html

14. http://a2zhomeschooling.com/wp-content/uploads/fig06.jpeg.492x0_q85_crop-smart.jpg

15. http://www.howitworksdaily.com/wp-content/uploads/2011/03/Tsunami-formation-large.jpg

16. http://weready.org/tsunami/images/Tsunami%20Smart%20-%20Causes%20-%20image%201.jpg

17. https://krissyhsie.files.wordpress.com/2010/03/00tectonicplatesx2.jpg

18. http://image.slidesharecdn.com/presentation1-141201131237-conversion-gate01/95/earthquake-20-638.jpg?cb=1429372028

19. http://image.slidesharecdn.com/earthquakes-150103040418-conversion-gate01/95/earthquakes-5-638.jpg?cb=1420257933

20. http://geology.isu.edu/wapi/envgeo/EG5_earthqks/seismogram.jpg

21. https://upload.wikimedia.org/wikipedia/commons/thumb/a/a3/Epicenter_Diagram.svg/2000px-Epicenter_Diagram.svg.png

22. http://images.slideplayer.com/15/4779089/slides/slide_25.jpg

23. http://image.slidesharecdn.com/earthquk-131215123315-phpapp01/95/earthquakes-16-638.jpg?cb=1387110891

24. http://legacy.earlham.edu/~kelleja/seiso1.gif

25. http://www.sleepingdogstudios.com/Network/Earth%20Science/ES_8.2_files/slide0016_image044.gif

26. http://www.kids-earth-science.com/images/eq6-determining-epicenter.png

Appendix 1: Earthquakes Tsunamis in India

1. *2004 Indian Ocean Earthquake and Tsunami*

https://en.wikipedia.org/wiki/2004_Indian_Ocean_earthquake_and_tsunami

Facts

Date	26 December 2004[1]
Origin time	00:58 UTC
Magnitude	9.1 M_w[1]
Depth	30 km (19 mi)[1]
Epicenter	3.316°N 95.854°ECoordinates: 3.316°N 95.854°E[1]
Type	Mega thrust
Areas affected	Indonesia Sri Lanka India Thailand Maldives Somalia
Tsunami	Yes
Casualties	230,000–280,000 dead and more missing[2][3][4][5]

Earthquakes and Tsunamis

The **2004 Indian Ocean earthquake** occurred at 00:58:53 UTC on 26 December with an epicenter off the west coast of Sumatra, Indonesia. The event is known by the scientific community as the Sumatra–Andaman earthquake.[6][7] The resulting tsunami was given various names, including the 2004 Indian Ocean tsunami, South Asian tsunami, Indonesian tsunami, the Christmas tsunami and the **Boxing Day tsunami**.

The undersea mega thrust earthquake was caused when the Indian Plate was sub ducted by the Burma Plate and triggered a series of devastating tsunamis along the coasts of most landmasses bordering the Indian Ocean, killing 230,000 people in 14 countries, and inundating coastal communities with waves up to 30 meters (100 ft) high. [8] It was one of the deadliest natural disasters in recorded history. Indonesia was the hardest-hit country, followed by Sri Lanka, India, and Thailand.

With a magnitude of M_w 9.1–9.3, it is the third-largest earthquake ever recorded on a seismograph. The earthquake had the longest duration of faulting ever observed, between 8.3 and 10 minutes. [9] It caused the entire planet to vibrate as much as 1 centimeter (0.4 inches) [10] and triggered other earthquakes as far away as Alaska.[11] Its epicenter was between Simeulue and mainland Indonesia.[12] The plight of the affected people and countries prompted a worldwide humanitarian response. In all, the worldwide community donated more than US$14 billion (2004) in humanitarian aid.[13]

References

1. Jump up to: [a][b][c][d][e] "Magnitude 9.1 – OFF THE WEST COAST OF SUMATRA". *U.S. Geological Survey*. Retrieved 26 August2012.

2. Jump up If the death toll in Myanmar was 400–600 as claimed by dissident groups there, rather than just 61 or 90, more than 230,000 people would have perished in total from the tsunami.

3. Jump up Earthquakes with 50,000 or More Deaths[dead link]

4. Jump up to: [a][b] "Indonesia quake toll jumps again". *BBC News. 25 January 2005*. Retrieved 24 December 2012.

5. Jump up Indian Ocean tsunami anniversary: Memorial events held 26 December 2014, BBC News

6. Jump up Lay, T., Kanamori, H., Ammon, C., Nettles, M., Ward, S., Aster, R., Beck, S., Bilek, S., Brudzinski, M., Butler, R., DeShon, H., Ekström, G., Satake, K., Sipkin, S., The Great Sumatra-Andaman Earthquake of 26 December 2004,Science, 308, 1127–1133, doi:10.1126/science.1112250, 2005

7. Jump up "Tsunamis and Earthquakes: Tsunami Generation from the 2004 Sumatra Earthquake -USGS Western Coastal and Marine Geology". *Walrus.wr.usgs.gov.* Retrieved 12 August 2010.

8. Jump up to: [a][b] *Paris, R.; Lavigne F., Wassimer P. & Sartohadi J. (2007). "Coastal sedimentation associated with the December 26, 2004 tsunami in Lhok Nga, west Banda Aceh (Sumatra, Indonesia)". Marine Geology (Elsevier) 238 (1–4): 93–106.* doi:10.1016/ j.margeo.2006.12.009.

9. Jump up NSF: "Analysis of the Sumatra-Andaman Earthquake Reveals Longest Fault Rupture Ever"

10. Jump up Walton, Marsha. "Scientists: Sumatra quake longest ever recorded." CNN. 20 May 2005

11. Jump up West, Michael; Sanches, John J.; McNutt, Stephen R. "Periodically Triggered Seismicity at Mount Wrangell, Alaska, After the Sumatra Earthquake." Science. Vol. 308, No. 5725, 1144–1146. 20 May 2005.

12. Jump up to: [a][b] Nalbant, S., Steacy, S., Sieh, K., Natawidjaja, D., and McCloskey, J. "Seismology: Earthquake risk on the Sunda trench." Nature. Vol. 435, No. 7043, 756–757. 9 June 2005. Retrieved 16 May 2009. Archived 18 May 2009.

13. Jump up Jayasuriya, Sisira and Peter McCawley, "The Asian Tsunami: Aid and Reconstruction after a Disaster". Cheltenham UK and Northampton MA USA: Edward Elgar, 2010.

2. *List of earthquakes in India*

https://en.wikipedia.org/wiki/List_of_earthquakes_in_India

The Indian subcontinent has a history of earthquakes. The reason for the high frequency and intensity of earthquakes is the Indian plate driving into Asia at a rate of approximately 49 mm/year. [1] The following is a list of **major earthquakes affecting India**.

Dates	Time	Location	Latitude	Longitude	Deaths (Overall)	Comments	Magnitude
October 26, 2015	09:09 UTC	Northern India, Pakistan, Afghanistan	36°14'45"N	71°50'38"E	260 in Pakistan and Afghanistan till 01.34 am on 27 Oct'15		7.7
June 28, 2015	06:35 IST	Dibrugarh, Assam	26.5°N	90.1°E	0	3 injured in Assam earthquake, tremors felt in West Bengal, Meghalaya and Bhutan	5.6
May 12, 2015	12:35 IST	Northern India, North East India	27.794°N	85.974°E	121+	Epicentre 17 km S of Kodari, Nepal; Felt in Delhi, West Bengal, Bihar, U.P.; 44 killed in India	7.3
April 26, 2015	12:39 IST	Northern India, North East India	27.794°N	85.974°E	Aftershock	Aftershock(Epicenter17 km S of Kodari, Nepal)	6.7[2]
April 25, 2015	12:19 IST	Northern India	28.193°N	84.865°E	Aftershock	Aftershock(Epicenter 49 km east of Lamjung, Nepal)	6.6[2]
April 25, 2015	11:41 IST	Northern India, North East India	28.147°N	84.708°E	8900+ [3]	Epicenter 34 km ESE of Lamjung,Nepal. Felt in eastern, northern, northeastern India and parts of Gujarat[4]	7.8[5]
March 21, 2014	18:41 IST	Andaman and Nicobar Islands	7.6°N	94.4°E	0	Moderate earthquake in Andaman Islands	6.7
April 25, 2012	08:45 IST	Andaman and Nicobar Islands	9.9°N	94.0°E	0	Big earthquake in Andaman and Niocbar Islands	6.2
March 5, 2012	13:10 IST	New Delhi	28.6°N	77.4°E	1	Moderate earthquake in national capital, CBSE Physics board exam disrupted in Delhi	5.2
September	18:10	Gangtok, Sikkim	27.723°N	88.064°E	118	Strong earthquake in NE	6.9

Date	Time	Place	Lat	Long	Deaths	Comments	Mag
18, 2011	IST	see 2011 Sikkim earthquake				India, tremors felt in Delhi, Kolkata, Lucknow and Jaipur	
August 10, 2009	01:21 IST	Andaman Islands see 2009 Andaman Islands earthquake	14.1°N	92.8°E	26	Tsunami Warning issued	7.7
October 8, 2005	08:50 IST	Kashmir see 2005 Kashmir earthquake	34.493°N	73.629°E	130,000	95 km (59 mi) NE of Islamabad, Pakistan, 125 km (78 mi) WNW of Srinagar, Kangra, Jammu and Kashmir, India(pop 894,000)	7.6
December 26, 2004	09:28 IST	off west coast northern Sumatra India Sri Lanka Maldives see 2004 Indian Ocean earthquake	3.30°N	95.87°E	283,106	Third deadliest earthquake in the history of the world, the tsunami generated killed 15,000 people in India	9.1
January 26, 2001	08:50 IST	Gujarat see Gujarat earthquake of 2001	23.6°N	69.8°E	20,000	Indian Republic Day Gujarat earthquake, thousands killed	7.6/7.7
March 29, 1999	00:35 IST	Chamoli district-Uttarakhand see 1999 Chamoli earthquake	30.408°N	79.416°E	103 Approx		6.8
August 21, 1988	04:40 IST	Udayapur, Nepal	26.755°N	86.616°E	~1000	6,553 injured	6.3–6.7
May 22, 1997	13:41 IST	Jabalpur, Madhya Pradesh	23.18°N	80.02°E	39		6.0
September 30, 1993	09:20 IST	Latur, Maharashtra see 1993 Latur earthquake	18.08°N	76.52°E	9,748		6.2
October 20, 1991	02:53 IST	Uttarkashi, Uttarakhand see 1991 Uttarkashi earthquake	30.73°N	78.45°E	>2,000		7.0
January 19, 1975	13:32 IST	Himachal Pradesh see 1975 Kinnaur earthquake	32.46°N	78.43°E	47		6.8
July 21, 1956	15:32 IST	Gujarat see 1956 Anjar earthquake	23.3°N	70.0°E	115		6.1
August 15, 1950	19:22 IST	Arunachal Pradesh see 1950 Assam–Tibet earthquake	28.5°N	96.7°E	1,526	Largest earthquake recorded in mainland India since Independence.	8.6
June 26,	08:50	Andaman Islands	12.50°N	92.57°E	7,000	Triggered a tsunami that	8.1

1941	IST	see 1941 Andaman Islands earthquake				affected eastern India and Sri Lanka	
May 31, 1935	03:02 IST	Quetta, Baluchistan see 1935 Balochistan earthquake	28.866°N	66.383°E	30,000 / 60,000	Deadliest earthquake recorded in the regions of modern-day Pakistan (then undivided India).	7.7
January 15, 1934	14:13 IST	Nepal see 1934 Nepal–Bihar earthquake	27.55°N	87.09°E	>10,000	Epicenter lies 10 km south of Mt. Everest.	8.0
April 4, 1905	01:19 IST	Kangra see 1905 Kangra earthquake	32.01°N	76.03°E	>20,000	It was a major earthquake that occurred in the Kangra Valley and the Kangra region of the Punjab Province (now Himachal Pradesh)	7.8
June 12, 1897	15:30 IST	Shillong, India see 1897 Assam earthquake	26°N	91°E	1,500		8.3
December 31, 1881	07:49 IST	Andaman Islands see 1881 Nicobar Islands earthquake	8.52 N	92.43 E	0	Earliest earthquake for which rupture parameters have been estimated instrumentally (from tide gauges)	7.9
June 16, 1819	18:45 to 18:50 local time	Gujarat see 1819 Rann of Kutch earthquake	23.0 N	71.0 E	>1,543	Formed the Allah Bund and Lake Sindri	8.2

References

1. "Earthquake Hazards and the Collision between India and Asia". Retrieved November 4, 2014.

2. "Current Month". Retrieved 27 April 2015.

3. Jason Burke. "Nepal earthquake death toll exceeds 6,000 with thousands unaccounted for". The Guardian.

4. "Nepal earthquake magnitude upgraded to 7.9, only 2-km deep: USGS". Times of india. indiatimes.com. 25 April 2015. Retrieved 25 April 2015.

5. "M7.8 - 34km ESE of Lamjung, Nepal". Retrieved 27 April 2015.

3. *Biggest earthquakes felt in India in the past 15 years*

http://timesofindia.indiatimes.com/india/Biggest-earthquakes-felt-in-India-in-the-past-15-years/articleshow/49538251.cms

A major earthquake that struck in the Hindu Kush region of Afghanistan around 2:40 pm IST caused massive tremors that were felt across most parts of northern India. It was estimated to have measured 7.5 on the Richter scale by the Indian Meteorological Department. Here is a list of major earthquakes that have been felt in India in the past 15 years:

1. Nepal earthquakes and aftershocks - Biggest quake on April 25, 2015 - 7.8 Tremors felt across India. About 9,000 killed in Nepal and India. May 12, 2015 - Nepal - 7.3 April 26, 2015 - Aftershock - 6.7 April 25, 2015 - Aftershock - 6.6 April 25, 2015 - Nepal - 7.8

2. April 25, 2012 - Andaman and Nicobar Islands - 6.2 None killed. Tsunami warning issued. Preventive evacuations in most coastal cities and towns.

3. Sikkim earthquake - September 18, 2011 - 6.9 Tremors felt across northern India. About 120 killed.

4. August 10, 2009 - Andaman Islands - 7.7 26 killed. Tsunami warning issued.

5. Kashmir earthquake - October 8, 2005 - 7.6 Tremors felt across India. About 1.3 lakh killed.

6. Indian Ocean earthquake - December 26, 2004 - Banda Aceh, Indonesia - 9.1 Third largest earthquakes ever recorded. Fourth most lethal natural disaster in history. Triggered tsunami that hit coasts of countries in Southeast Asia, South Asia, Australia and Africa. About 2.8 lakh killed. Of about 11,000 in India, a majority were in Tamil Nadu.

7. Gujarat earthquake - January 26, 2001 - 7.7 Quake hit near Bhuj in Kachchh, flattening many towns and cities in Gujarat. Over 20,000 killed.

Warning / Comment: *One should remain alert, if the main centre exists in Himalaya range then it will damage maximum part of northern India*

4. *India Disaster Knowledge Network*

http://www.saarc-sadkn.org/countries/india/hazard_profile.aspx

Hazard Profile

India, due to its, physiographic and climatic conditions is one of the most disaster prone areas of the world. It is vulnerable to windstorms from both the Arabian Sea and Bay of Bengal. There are active crustal movements in the Himalaya leading to earthquakes. About 58.7 % of the total land mass is prone to earthquake of moderate to very high intensity.

The region was hit by Uttarkashi Earthquake(1991), Killari Earthquake(1993), Koyana Earthquake(1997),Chamoli Earthquake(1999), and Bhuj earthquake (2001), Jammu&Kashmir Earthquake (2005). The Himalayas being a fairly young mountain range is undergoing constant geological changes resulting in landslides. Floods brought about by heavy rain and drought in arid and semi arid areas. About 12 % of the o f the total land mass is flood prone and 68 % of the arable land is vulnerable to drought. The Western region of the country is represented by the Thar Desert and the central India by the Deccan Plateau face recurring droughts due to acute shortage of rainfall.

India has increasingly become vulnerable to Tsunamis since the 2004 Indian Ocean Tsunami. India has a coastline running 7600 km long; as a result is repeatedly threatened by cyclones.

5. Major Disasters in India

The Table below shows major disasters in the known history of India

SR. NO.	Name of Event	Year	Fatalities
1.	Maharashtra Earthquake	1618	2,000
2.	Bengal Earthquake	1737	300,000
3.	Bengal Cyclone	1864	60,000
4.	The Great Famine of Southern India	1876-1878	5.5 million
5.	Maharashtra Cyclone	1882	100,000
6.	The Great Indian famine	1896-1897	1.25 million to 10 million
7.	Kangra earthquake	1905	20,000
8.	Bihar Earthquake	1934	6,000
9.	Bengal Cyclone	1970	500,000 (include Pakistan and Bangladesh also)
10.	Drought	1972	200 million people affected
11.	Andhra Pradesh Cyclone	1977	10,000
12.	Drought in Haryana & Punjab	1987	300 million people affected
13.	Latur Earthquake	1993	7,928 death and 30,000 injured
14.	Orissa Super Cyclone	1999	10,000
15.	Gujarat Earthquake	2001	25,000
16.	Indian Ocean Tsunami	2004	10,749 deaths 5,640 persons missing
17.	Kashmir Earthquake	2005	86000 deaths (include Kashmir & Pakistan)
18.	Kosi Floods	2008	527
19.	Cyclone Nisha of Tamil Nadu	2008	204

6. 10 Earthquake Prone Cities in India

http://indiatoday.intoday.in/education/story/top-10-earthquake-prone-cities-in-india/1/432331.html

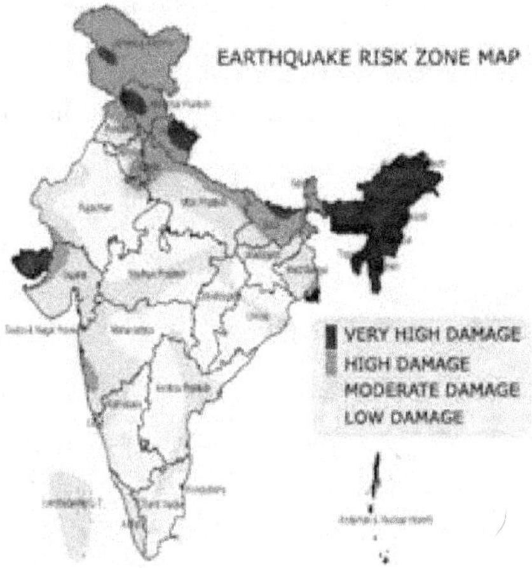

Earthquake Prone Cities in India

The Indian subcontinent has a history of devastating earthquakes. An earthquake of 7.2 magnitude struck Tajikistan and the tremors were felt in India and Pakistan.

The epicenter of the earthquake was Tajikistan, measuring 7.2 on the Richter Scale and was recorded at 25 km.

India is very prone to earthquakes as well. The major reason for the high frequency and intensity of the earthquakes is that the Indian plate is driving into Asia at a rate of approximately 47 mm/year. As per the Geographical statistics, almost 54% of the land in India is vulnerable to earthquakes.

According to the estimates shown by a World Bank and United Nations report; around 200 million city dwellers in India will be exposed to storms and earthquakes by 2050.

The latest version of the seismic zoning map of India assigns four levels of seismicity for India in terms of zone factors, which means India is divided into 4 seismic zones:

- Zone 2
- Zone 3
- Zone 4
- Zone 5

Zone 5 is highly prone to the earthquake with the highest level of seismicity whereas Zone 2 is associated with the lowest level of seismicity. So, the Zones - marked two to five -indicate areas most likely to experience tremors with zone five being the most vulnerable.

Indian cities, ranging from the metros to the smaller cities - all at least once have been shaken up due to earthquakes which usually range from medium to high intensity on the Richter scale.

The top 10 Indian cities which are observed as high earthquake prone zones are shown below:

1. *Guwahati - Assam*

Guwahati falls in zone five of the seismic zones in India which is highly prone to earthquakes. The place has seen some of the deadliest earthquakes and even today small tremors are a common situation. Guwahati receives earthquake predictions on a daily basis; resulting which many adjoining areas in the North-East get affected.

2. *Srinagar - Jammu and Kashmir*

This capital city of Jammu and Kashmir also comes under Seismic Zone 5.

Most parts of the Kashmir Valley, which is around 11% of the area of the state covering the Districts of Srinagar, Ganderbal, Baramulla, Kupwara, Bandipora, Budgam, Anantnag, Pulwama, Doda, Ramban, Kishtwar, come under Seismic Zone 5, where around 50% of the population of the state lives. The rest of the state, including the whole of the Ladakh region and Jammu Division (90% of the total area of the state), are under the Seismic Zone 4.

Being very close to the Himalayas, Srinagar faces heavy risk of earthquakes, high as well as moderate. The friction between the Indian and the Eurasian plane causes earthquakes to occur on the areas close to the Himalayas.

3. *Delhi*

Delhi is categorized under Seismic Zone 4. Delhi has been hit by five devastating earthquakes measuring higher than magnitude of 5 since 1720. The most prone to earthquake neighbourhoods in Delhi lie about two miles on either side of the Yamuna river, the southwestern outskirts of the city known as the Chhattarpur basin, as well as an area popularly known as The Ridge in Delhi

4. *Mumbai - Maharashtra*

Mumbai falls in the Zone 4 of the seismic zone divisions which makes it quite vulnerable to damage.

We all know Mumbai is located on the coastal line, which increases the risk of facing tsunami-like disasters. Mild to strong earthquakes are very common in parts of Mumbai. Mild earthquakes are often faced by people living there and parts of the adjoining regions of Gujarat. It should be noted that for the last 20 years, almost all of the buildings in Mumbai have been designed and built keeping in mind that the city falls in seismic zone 4.

5. *Chennai - Tamil Nadu*

The city, formerly in the comfort Zone 2, has now shifted to Zone 3 - indicating higher seismic activity. According to the seismic mapping , districts in the western part along the border with Kerala are also in Zone 3, along with districts along the border of Andhra Pradesh and a section of the border with Karnataka.

The status of Chennai along with major towns on the eastern coast in terms of vulnerability has increased especially after Chennai experienced tremors in September 2001 following a quake measuring 5.6 on the Richter scale off the Pondicherry coast.

Tamil Nadu, had faced the wrath of the deadly 2004 tsunami when the Marina beach was affected.

Recently, in the year 2012, Chennai shook terribly due to a rather high intensity earthquake (having its epicentre in the Indian Ocean).

The other 5 cities in right order of chronology are:

6. *Pune - Maharashtra*

7. *Kochi - Kerala*

8. *Kolkata - West Bengal*

9. *Thiruvananthapuram - Kerala*

10. *Patna – Bihar*

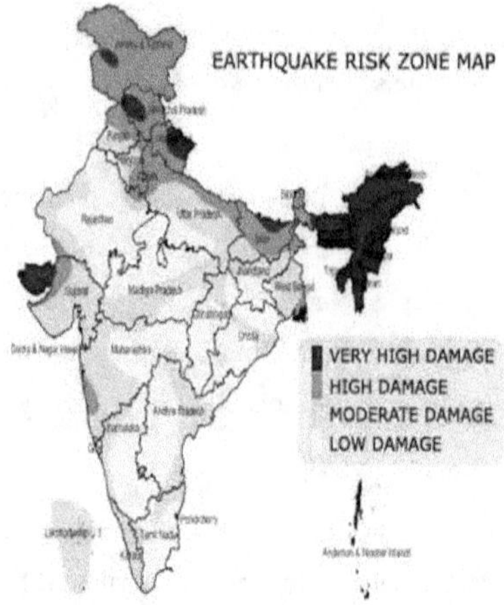

(Source: National Disaster Management Authority. GOI Image Source: NDMA)

Appendix 2: All About Earthquakes: Utah Earthquakes

1. Utah Earthquake History

-July 18, 1894: Ogden, VI-VII (Mag 5.0) -August 1, 1900: Santaquin, VI-VII -November 13, 1901: Parowan-Richfield, VIII

-November 14, 1901: Parowan-Richfield Aftershocks, VII -November 17, 1902: Pine Valley, St. George, Santa Clara, VII

-October-December 1909: Series of 30-60 quakes within Garland and Tremonton, VII -May 22, 1910: Salt Lake City, VII (Mag 5.0) -May 13, 1914: Ogden, VII (Mag 5.5)

-September 29, 1921: Elsinore, Monroe, Richfield, 2 strong earthquakes 12 hours apart, VIII -March 12, 1934: Kosmo, shore of Great Salt Lake, VIII (Mag 6.6)

-August 30, 1962: Franklin, Logan, Preston, Richmond (Cache County) Mag 5.7 -October 4, 1967: Marysvale, Mag 5.2

-March 28, 1975: Idaho-Utah border, Mag 6.1 **See more info at:[1]

2. Largest Earthquake in Utah

-March 12, 1934: Hansel Valley, near Kosmo (30 miles north of Great Salt Lake) Mag 6.6 Newspaper Articles [2]

3. Salt Lake City Earthquake -May 22, 1910: Salt Lake City, Mag 5.5 Newspaper Articles [3]

4. Wasatch Fault Zone, Salt Lake City

The Wasatch fault zone is one of the longest and most tectonically active normal faults in North America. The fault zone shows abundant evidence of recurrent Holocene surface faulting and has been the subject of detailed studies for over three decades. Half of the estimated 50 to 120 post-Bonneville surface-faulting earthquakes in the Wasatch Front region have been on the Wasatch fault zone.

This fault has 10 sections. The nearly 350-km-long Wasatch fault zone has traditionally been divided into seismogenic segments that are thought to rupture at least somewhat independently. The southern eight sections are entirely in Utah. To the north, the Clarkston Mountain section straddles the state line between Idaho and Utah and the northernmost (Malad City) section is entirely in Idaho. The chronology of surface-faulting earthquakes on the Wasatch fault is one of the best dated chronologies in the world and includes 16 earthquakes since 5.6 ka, with an average repeat time of 350 yr. Four of the central five sections [2351e-h] ruptured in the last hundreds to about a thousand years ago, whereas the next section to the north, Brigham City [2351d], has not ruptured in the past 2,125 yr. Slip rates of 1-2 mm/yr are typical for the central sections during Holocene time. In contrast, middle and late Quaternary (<150-250 ka) slip rates on these sections are about an order of magnitude lower. This substantial change in the slip rate may indicate a causal relation between increased Holocene slip rates and isostatic rebound/crustal relaxation following deep lake cycles such as Bonneville [4]

Where: Davis County, Salt Lake County

Land Features: Basin and Range

Length: This section is 43 km of a total fault length of 357 km

Slip rate: Between 1.0-5.0 mm/year

References

1. http://earthquake.usgs.gov/earthquakes/states/utah/history.php
2. http://www.seis.utah.edu/lqthreat/nehrp_htm/1934hans/n1934ha1
3. http://www.seis.utah.edu/lqthreat/nehrp_htm/1910salt/n1910sl1.shtml#tessbslhttp://geohazards.usgs.gov/cfusion/qfault/qf_web_disp.cfm?disp_cd=B&qfault_or=5326&ims_c f_cd=cf
4. http://geohazards.usgs.gov/cfusion/qfault/qf_web_disp.cfm?disp_cd=B&qfault_or=5326&ims_c f_cd=cf

www.ingramcontent.com/pod-product-compliance
Lightning Source LLC
Chambersburg PA
CBHW071951150726
47999CB00001B/398